COMPUTER-INTEGRATED MANUFACTURING:

The social dimension

Karl-H. Ebel

International Labour Office   Geneva

ISBN 92-2-107276-2

First published 1990

Cover photograph: ILO Photo Library, Geneva/Renault. A finishing robot welding together parts in a car factory workshop.

Printed in Switzerland   ATA

D
621.78
EBE

DmCC

FOREWORD

The debate on the implementation of computer-integrated manufacturing (CIM) has made the headlines in the past few years. It has centred primarily on the technical and economic feasibility of this new manufacturing concept and on the prospects of raising productivity and making enterprises more competitive. The social implications of CIM are only beginning to be explored, mainly because actual realisations are few and far between. However, this new technology has the potential for changing the world of work profoundly. It is, therefore, opportune for the ILO to assess the possible social and labour effects according to the available evidence.

This study reflects the great divergence of methods and views world-wide. It argues that the approach to CIM that has the best chance of succeeding is one that takes the human factor fully into account. Furthermore, there is no reason to be unduly alarmed about a manufacturing revolution: changes will be incremental and the "unmanned factory" is not in sight. The study will, it is hoped, stimulate the search for technical and organisational forms of CIM that will improve productivity and the working environment and that are compatible with people's striving for satisfying and motivating work and working conditions. It is also intended to give the social partners in collective bargaining an overview of the current state of CIM and the issues that they may have to address in the future.

Jürgen von Muralt
Director,
Sectoral Activities Department

ACKNOWLEDGEMENTS

This comparative study owes much to the interchange of ideas and of research results which took place on the occasion of the three ILO workshops organised in 1987, 1988 and 1989 under the auspices of the ILO/Federal Republic of Germany Project on Expert Systems and Qualification Changes; the CIM workshops organised in 1987 and 1989 by the International Institute for Applied Systems Analysis (IIASA), Laxenburg (Austria); the United Nations Economic Commission for Europe Seminar on Computer-Integrated Manufacturing, Botevgrad (Bulgaria), September 1989; and the International Federation of Automatic Control (IFAC) Symposium on Skill-based Automated Production, Vienna (Austria), November 1989.

Jean-Pierre Durand, Université de Rouen (France); Paul Kidd, Cheshire Henbury Research and Consultancy (United Kingdom); Erhard Ulrich, Institut für Arbeitsmarkt- und Berufsforschung der Bundesanstalt für Arbeit, Nuremberg (Federal Republic of Germany); and Werner Sengenberger, International Institute for Labour Studies (ILO, Geneva, Switzerland) commented and made much appreciated suggestions. The author also thanks his ILO colleagues, Gabriele Stoikov, Jon McLin and Claudio de Moura Castro, for devoting time to discuss the issues, but takes responsibility for the assessment of trends and opinions expressed.

CONTENTS

LIST OF TABLES AND FIGURES

ABBREVIATIONS IN THE FIELD OF CIM

AGV Automated guided vehicle
AI Artificial intelligence
AMH Automated material(s) handling
AMT Advanced manufacturing technology
ATE Automatic text equipment
CAA Computer-aided assembly
CAD Computer-aided design
CAD/CAM Computer-aided design and manufacturing
CADD Computer-aided design and drafting
CAE Computer-aided engineering
CAI Computer-aided inspection
CAM Computer-aided manufacturing
CAP Computer-aided planning
CAPM Computer-aided production management
CAPP Computer-aided process planning
CAQ Computer-aided quality assurance control
CAR Computer-aided robotics
CAS Computer-aided simulation or computer-aided software engineering
CAT Computer-aided testing
CIM Computer-integrated manufacturing
CNC Computerised numerical control
DNC Distributive numerical control
EDP Electronic data processing
FMS Flexible manufacturing system
IT Information technology
JIT Just in time
LAN Local area network

MAP Manufacturing automation protocol
MPCS Manufacturing planning and control system
MRP Manufacturing resource planning
NC Numerical control
OPT Optimised production technology
PA Programmable automation
PC Personal computer
TOP Technical and office protocol
TQC Total quality control
VDU Visual display unit

INTRODUCTION

There is a growing body of opinion, supported by case studies and research, that industry is at the threshold of a new era in manufacturing. Many believe that computer-integrated manufacturing (CIM) will transform the world of work beyond recognition. However, there is also much evidence and many cautionary tales of failures following the introduction of this new technology – although failure is seldom advertised because it affects a company's image. Has the CIM concept the potential for introducing far-reaching qualitative changes in manufacturing? And what can we reasonably expect to happen to the men and women employed in manufacturing? A preliminary answer to these questions, emphasising the key role of the human factor in the application of the new technology, is attempted in this study.

There is no generally accepted definition of CIM. Some observers take a narrow view and regard CIM as the linking of computer-aided design (CAD) and computer-aided manufacturing (CAM). Others take a broader view and see it partly as a methodology involving a consistent systems-based approach to running a manufacturing business, and partly as the integration of information and the integration of business and manufacturing activities. The CIM concept therefore starts with integration of the various aspects of production (e.g. design, planning, manufacturing, warehousing, materials handling and quality control), but may also incorporate business aims and objectives, sales and marketing, and financial management. In some cases it is concerned with the linking of geographically separate sites. It can also involve electronically linking the business with its customers and suppliers.

The functions and subfunctions in each area are parts of a system, are fully integrated through computer networks and have access to a unified data base. CIM is thus essentially a means of organising and controlling the manufacture of components and assemblies as logically and flexibly as possible and of mastering and co-ordinating the corresponding flow of data and information. It aims at optimising the use of equipment, decreasing lead time and stocks, and ensuring high product quality and lower unit costs. The synergies created through integration are expected to lead to cost reductions, higher productivity, and rapid adjustment of product quantity and quality and product variations, as well as delivery times to demand in competitive national and international markets. It appears to offer an opportunity to keep up with shorter product life cycles and to eliminate a good deal of waste.

The CIM concept seems rational and sensible, and appeals to the tidy mind. It satisfies the quest of the engineer to create order out of chaos. It assures and comforts the manager looking for efficient means of controlling the production process. Why is it then that the realisation of projects runs into so much difficulty in practice? Are expectations too high? Even allowing for the fact that CIM can be introduced only step by step, that it requires considerable computing power and the mastery of complex system architectures and software developments, and that different production processes need varying CIM systems (i.e. they must be tailor-made), it is by now obvious that the practical results achieved so far have not lived up to the initial optimism of many automation equipment and system suppliers, engineering researchers and management strategists. Indeed, in the United Kingdom international management consultants AT Kearney have raised concerns about the overall direction and effectiveness of CIM applications. The conclusions of their recent survey of British companies indicate that CIM has not resolved the problems of quality and performance to schedules as anticipated.[1]

The "factory of the future", therefore, as described by CIM advocates in the engineering profession, remains largely a figment of the imagination despite a few science-fiction type realisations. And even staunch technocrats readily admit that industry is a long way from realising the full potential of CIM, although partial solutions such as CAD, flexible manufacturing systems (FMS), automatic materials handling (AMH) and computer-numerical control of machine tools (CNC) have made promising headway. "Islands of automation" have thus been created in many plants, but linking them is clearly not an easy task.

Some CIM systems have been set up as demonstration or pilot projects for this new approach to manufacturing, but they are usually very costly and isolated experiments outside the real world of production. Already in 1986 at Leeds University in the United Kingdom, 60 vendors including IBM and Digital Equipment, the leaders in factory automation, set up the CIMAP demonstration system linking technical office, design and engineering, machinery, robotics, quality assurance, assembly and inspection through standardised networks and protocols, i.e. the manufacturing automation protocol (MAP) and the technical and office protocol (TOP). The subsystems were proprietary and incompatible but could communicate with each other through the standard protocols.[2]

Another example of a pilot project, IMPACT, was shown at the 1989 Advanced Manufacturing Systems Exhibition and Conference, Genoa, Italy. IMPACT is a mini-factory designed to demonstrate the integration of materials handling, production and control technology. Arthur Andersen Consultants co-ordinated the 16 firms

[1] AT Kearney Management Consultants: *Computer-integrated manufacturing: Competitive advantage or technological dead end?* (London, 1989).

[2] M. Kangas: *Computer-integrated manufacturing: A new manufacturing concept*, Report No. 005 (Geneva, International Management Institute, Dec. 1987), pp. 58 ff.

from various countries which co-operated in this project. In previous years a similar mini-factory was exhibited at the Advanced Manufacturing Systems Fair in Chicago.[3]

On the whole, the installation of operational CIM systems in industry is in inverse proportion to the talk about it. Should we, therefore, regard CIM as a blind alley, a passing fad or a technocrat's pipe-dream of the manufacturing paradise?

As experience in this field accumulates through trial and error as well as through failures and successes, and as different schools of thought try to impose their particular vision of the "factory of the future", it becomes increasingly clear that there are various dimensions to the problem — socio-economic, technical, managerial and human — that are partly rooted in industrial history.

[3] "'Co-operative' CIM at Genoa AMS show", in *American Machinist* (New York), May 1989, p. 35.

THE SOCIO-ECONOMIC PERSPECTIVE

The CIM concept has evolved in the highly industrialised countries characterised by significant capital accumulation, high labour costs, a broad and solid scientific and technological base, and a well-developed social and economic infrastructure. It owes its birth to rapid advances in computer and information technology. Pilot projects and accompanying research to put the concept into practice in manufacturing are concentrated in Japan, the United States and some industrially advanced European countries. However, there are considerable differences in the approaches chosen, which tend to be a response to the specific socio-economic situation, the industrial traditions and the factor endowment of the country or enterprise concerned.

The human-centred versus the technocentric approach

These different approaches may be broken down essentially into two: "technocentric" and "human centred".[1] No industrial society has the monopoly of one or the other approach and they

[1] P. Brödner: "Humane work design for man-machine systems — A challenge to engineers and labour scientists", in *Proceedings of the IFAC Conference on Analysis, Design and Evaluation of Man-Machine Systems,* Baden-Baden, 1982.

frequently coexist, although one may tend to predominate. The following analysis of the main features of these approaches, which does not go into their historical roots, should be seen in this perspective.

In the United States the so-called "technocentric" approach is found in its purest form. It has often served as a model for enterprises in other countries.[2] It denotes an attempt to gradually reduce human intervention in the production process to a minimum and to design systems flexible enough to react rapidly to changing market demand for high quality products. Workers and technicians on the shop-floor are sometimes seen as unpredictable, troublesome and unreliable elements capable of disturbing the production and information flow which is best controlled centrally through computers. The "unmanned factory" is the ultimate goal. It represents the division of labour carried to its extreme whereby subdivided and simplified tasks executed by a mass of low-skilled labour are progressively taken over by increasingly flexible intelligent and versatile industrial robots and machine systems communicating among each other via networks and computers.

Only a residual role is assigned to workers, whose skills are supposed to be gradually incorporated into and progressively embodied in the machines. The technocentric approach, it is hoped, will halt the continuing erosion of American production know-how and help industry to regain its lost superiority and competitivity in world markets. Investment in capital-intensive and sophisticated technology (some have dubbed this the "moonshot" approach) is thus expected to overcome a deep-seated structural problem. Since American manufacturers produce for a vast and homogeneous home market, they can concentrate on high-volume products and fairly large batches in component manufacturing. The resulting production

---

[2] W. Wobbe: "Technology, work and employment – New trends in the structural change of society", in *Vocational Training Bulletin* (Berlin (West), European Centre for the Development of Vocational Training CEDEFPOP), 1987, No. 1, pp. 3-6; P. Brödner: "Towards an anthropocentric approach in European manufacturing", ibid., pp. 30-39.

process is relatively inflexible even when flexible manufacturing systems and machining cells are used. The central engineering challenge is said to consist of arriving at continuous flow production of large varieties of products and components without much work-in-process, i.e. without idling capital; this would ensure high productivity and adequate returns. Human skills play a minor role in this scenario, which is a late vindication of Taylorism or Fordism, originally based on the principle of using vast pools of unskilled and semi-skilled labour. Today increasingly sophisticated machinery is designed to do away with jobs that have first been fragmented and de-skilled. Traditionally adversarial industrial relations and low commitment and loyalty on the part of the workforce reinforce this attitude. The technical office staffed by professional engineers and technicians increasingly becomes the repository of production know-how to the detriment of production workers.

Carried to its extreme, this approach has proved not to work very well or functions satisfactorily only at excessively high cost. For instance, it has been found that flexible manufacturing systems installed in the United States often performed worse than conventional technology. It is significant that America's biggest machine-tool manufacturer has announced the closure of its centralised FMS facility in favour of smaller manufacturing cells; other American machine-tool companies have also confirmed that they are moving away from large and highly complex FMS installations.[3] The relevance of the technocentric approach for the future of manufacturing therefore seems questionable.[4] It may prove to be a dead-end because of an essential flaw: there is mounting evidence that proper and continuous operation of the type of flexible automation that is central to CIM systems can be ensured only by highly qualified and motivated workers able to cope with the

---

[3] J. Dunn: "Flexible friends no more", in *The Engineer* (London), 15 Sep. 1988, p. 34.

[4] S. S. Cohen and J. Zysman: "US competitiveness suffers: The emergence of a manufacturing gap", in *Transatlantic Perspectives* (Washington, DC), No. 18, Autumn 1988, pp. 6-9.

relatively frequent breakdowns of such complex and sophisticated equipment, and with software problems — and there are persistent complaints that the specific skills needed for "high-tech" manufacturing are scarce or simply not available. If there is a lack of, or a shortage of, committed, highly skilled and well-trained staff, systems tend to perform far below their potential capacity. Inadequate performance of such sophisticated installations appears to be the rule rather than the exception. At any rate, there is generally a long learning and running-in period, with uncertain future returns as a result of excessive reliance on unproven technology.

The Japanese approach to CIM is based on a different rationale. Companies introducing advanced and flexible automation systems can rely on a highly qualified, versatile and loyal workforce. Instead of progressing in the technological field by giant leaps they prefer to make gradual improvements in the production process and in quality, often initiated by motivated engineers, technicians and workers on the shop-floor. The wide adoption of the quality circle movement is only one manifestation of this. Emphasis is on close co-ordination of design and process engineering, product quality and production scheduling (i.e. just-in-time production and electronic ordering of materials and components). Compared with the outstanding manufacturing skills evident in the wide application of industrial robots, the integration of the information flow through computers (i.e. the application of information technology) is relatively less developed due to gaps and difficulties in software development. The most flexible element in the system is, in fact, the people who make it work. Moreover, most companies operating such systems tend to serve large local and export markets and therefore produce relatively large series, although the flexibility of the equipment is more fully utilised than in American manufacturing, thanks to the highly qualified workforce. The strength of this approach, which is facilitated by a co-operative industrial relations system, is apparent in diversified high-quality mass production.

Manufacturers catering for relatively small, heterogeneous or specialised internal or export markets demanding high-quality

Computer-integrated manufacturing

components and customised products, as is largely the case in Europe, have been inclined to rely on another strategy in the introduction of CIM. In view of the high investments required and the often limited capital base, they usually opt for a cautious, pragmatic and gradual approach rather than adopting the whole panoply of CIM at once. The centrepiece of manufacturing has remained, by and large, the skilled and highly skilled worker or technician. While Taylorism made some inroads in European manufacturing, particularly in the automobile and consumer durables industries, it never replaced skill-based production in medium and small-scale enterprises in the capital goods sector, where an extreme division of labour is not feasible. Enterprises competing in narrow markets or occupying market niches have always had to be flexible and innovative to survive. The new computerised flexible and integrated automation equipment is seen primarily as an improved tool in the hands of a skilled and versatile workforce serving to enhance existing know-how, and to permit greater flexibility, higher productivity, better product quality and shorter delivery times. It is regarded not as a panacea for all production problems, but as an effective prop in gaining a market share. Such enterprises also tend to make a sustained effort to retrain their staff. Moreover, the lesser emphasis on division of work allows them to assign broader responsibilities to workers according to their qualifications, and consequently permits more flexible forms of managerial control and organisation, including team-work and imaginative applications of CIM using available skills.[5] The following examples reported by Caulkin may serve to illustrate the point made above:

Consider two examples of high technology implementation. First, GM's car plant at Lordstown, Ohio. Opened with fanfares in 1972, this brand new automated plant was an expensive disaster. It suffered from the start from absenteeism, high labour turnover, disappointing productivity and all the other signs of chronic worker alienation. Now look at Nissan's plant in the UK. The

---

[5] H.-J. Warnecke: *CIM in Europe* (Stuttgart, 1987), unpublished manuscript.

first Japanese auto plant in Europe, it is also highly automated and quite different in structure and organisation from any other car assembly site in the UK. Nissan Washington has met all its quality and productivity targets and with a $1,000 cost advantage per car is reportedly the lowest cost car plant in Europe. In 1988 it decided to go ahead with a second phase of expansion and recruitment.

Lordstown was a classic case of seduction by technology. In designing the plant for machines instead of humans, GM, like many other manufacturers, missed the opportunity to turn the workforce from part of the manufacturing problem (the implicit Western assumption) into part of the solution. Nissan is the reverse, an object lesson which demonstrates that advanced IT-based manufacturing has to integrate the human factor as much as the technology itself. As long as there are manned plants, there are no exceptions to this rule.

In retrospect, it seems clear that the first wave of CIM-type ... automation was unsuccessful partly because it was simply the extreme form of the manufacturing obsession with reducing direct labour costs. Behind it was the implicit notion that getting rid of people was necessarily a good thing. It is equally clear that in a general manner one of the most important reasons for the present impasse in Western manufacturing is the failure to make proper use of the human element. The new manufacturing, on the other hand, offers a unique opportunity to recombine the jobs fragmented by previous well-intentioned job design and for the first time unite the benefits of both men and machines.[6]

Economic benefits

Only rough estimates of the potential economic benefits of CIM are available. Normally based on only a few cases, they are produced by equipment vendors or are the result of research into limited applications. They are unreliable in so far as current business accounting methods do not permit the calculation of the real returns to such investment — or at best do so only approximately. The following indications are therefore given with due reservation and should serve merely as a basic orientation.

Provided CIM introduction succeeds, it is expected to lead to a 10 per cent rise in output; a 5-20 per cent reduction in personnel costs; a 10-15 per cent reduction in production costs; a 20 per cent reduction in stocks; a 10 per cent reduction in scrap; and order

[6] S. Caulkin and Ingersoll Engineers: "The human factor in IT: Man and machine", in *Multinational Business* (London, Economist Intelligence Unit), No. 1, 1989, p. 2.

lead-time reductions of about 50 per cent.[7] Another estimate assumes the potential overall cost reduction to be 5-10 per cent, i.e. more than the profit margin of most companies, and points out that CIM mainly serves to eliminate "hidden" costs and overheads, such as the accumulation of work-in-process and idle machine time.[8] More indirect benefits, which help to maintain or improve the market position of enterprises, include shorter product development cycles, the accelerated production of prototypes and reduced warranty and after-sales service costs. The time-span within which such economic benefits could materialise is not specified.

In 1984 the Manufacturing Studies Board of the National Research Council of the United States arrived at the most optimistic figures: 5-20 per cent reduction in personnel costs; 15-30 per cent reduction in engineering costs; 30-60 per cent reduction in overall lead-time; 30-60 per cent reduction in work-in-process; 40-70 per cent increase in overall production; 200-300 per cent gain in equipment operating time; 200-500 per cent increase in product quality; and 300-3,500 per cent gain in engineering productivity.[9]

All investigations tend to concur that the actual saving in labour costs is marginal considering that these normally range between 5 and 15 per cent of the total product cost. CIM simply means that there will be less, but better-remunerated highly skilled labour. Such savings do not offset the higher capital cost of introducing CIM. However, it has been suggested that the reduction of excess stocks and work-in-process could potentially finance all reasonable investment in CIM.[10]

In more general terms it is significant that the CIM concept implies a shift from economies of scale typical of mass production to

[7] G. Garner: "Potentials of CIM", in *Chemical Engineering* (New York), 28 Mar. 1988.

[8] Warnecke, op. cit.

[9] Kangas, op. cit., p. 33.

[10] "Factory of the future — A survey", in *The Economist* (London), 30 May 1987, pp. 17-18.

economies of scope. Many product designs and variations can be manufactured cost-effectively irrespective of quantity. This greater product variety improves the competitive position of enterprises in fragmented markets in that costs are spread over a wide variety of output, whereas in mass production costs are spread over as many identical units as possible. The shift occurs gradually, but the trend is already clearly visible, for instance, in automobile manufacturing and consumer electronics.

Beyond the narrow confines of the enterprise, CIM will have repercussions on the economy and society, as well as on the environment, which are difficult to assess at the present stage. On the one hand, CIM can mean more economic use of energy and materials, higher quality and, therefore, greater durability of products. On the other, it may also reduce the life-cycle of products by making them rapidly obsolescent and thereby inducing consumers to go for the latest novelties. This will have implications for raw materials and energy consumption and the recycling of waste, possibly at considerable cost to society as new materials for which no safe waste disposal has been developed are constantly used. The total benefit to society from CIM will finally depend on the options taken by enterprises, consumers and the government. Industry might well be asked to contribute to the elimination of detrimental effects. IBM's recently announced programme for the recycling of obsolete computers may become a model. The increasing amount of plastic materials used in cars is likely to require new methods of recycling as well.

Moreover, CIM technologies and in particular telematics, allow decentralisation of production, e.g. by centralising design at headquarters and locating manufacturing in dispersed small plants. This is bound to have an environmental impact which is difficult to foresee at present.

The diffusion of CIM

As few CIM systems are in operation and only partial realisations of the concept exist, it is only possible to assess the diffusion of subsystems such as CAD, CAM, CNC and FMS. From studies undertaken in this field,[11] one is led to conclude that fundamental changes are taking place in the approach to manufacturing, notably a move towards flexibility, but that in the real world changes proceed along an evolutionary path. Although the recent investment boom in the industrialised world has speeded up modernisation to some extent, CIM has certainly not become a dominant feature. The density of use of these new techniques was found to be particularly high in Japan and Sweden, but surprisingly low in the United States. Moreover, while there was a rising trend in installations in most industrialised countries as well as in newly industrialised countries (NICs), the United States did not keep up with this trend.

Since the United States continues to take the lead in science and technology in most fields and has set high standards of manufacturing productivity in the past, the threat of being left behind in competitive manufacturing technology has provoked searching questions and some explanations. The facts are these: in 1987 more than half American metalworking plants did not even have one computerised machine. Whereas only 11 per cent of the stock of machine tools in the United States was computer controlled, this was already the case for about 30 per cent of them in Japan in 1985. Small and medium-sized firms in the United States were found to be particularly reluctant and ill-equipped to adopt best practice technologies and did not perceive much economic advantage because

[11] C. Edquist and S. Jacobsson: *Flexible automation – The global diffusion of new technology in the engineering industry* (Oxford and New York, Basil Blackwell, 1988); International Institute for Applied Systems Analysis (IIASA), CIM study, publication forthcoming; *Automatisation et nouvelle gestion du travail: Réflexions sur la CFAO*, Etudes et Recherches de l'ISÈRES, No. 63, Montreuil, 1989).

of uncertain pay-offs. This technical stagnation in the small manufacturing sector was leading to a loss of overall innovative capacity, which meant that component supplier industries were losing out to foreign competition. On the whole, technological leaders in the United States stayed with rigid dedicated automation and placed much more emphasis on product innovation than on process innovation. The latter, however, produces more rapid results in terms of increased market shares and employment gains.[12]

A sample survey carried out in the Federal Republic of Germany in 1987-88 gives a fairly detailed insight into the diffusion of CIM technologies in the capital goods industry of that country. They are not yet very widely used: while about half of all enterprises use CNC machines — a relatively mature technology — only 3.7 per cent are equipped with flexible manufacturing cells (FMC); 2.8 per cent with FMS; 3.4 per cent with distributive numerical control (DNC) systems; 8.6 per cent with industrial robots; 5.1 per cent with computer-aided assembly (CAA) systems; 4.8 per cent with automatic warehousing; and 0.7 per cent with AMH. Between 15 and 17 per cent of capital goods manufacturers use computer-aided process planning (CAPP), CAM and CAD and about 8 per cent use computer-aided quality assurance (CAQ). The survey predicted that by 1990 the number of enterprises using CAPP, CAM and CAD would double. As expected, large enterprises make most use of CIM components. There is, however, a sizeable proportion of medium-sized enterprises (100-500 employees) and even of smaller firms that have begun to invest in these systems or are planning to do so.[13]

There are, of course, many obstacles to faster diffusion of computerised technology, as will be shown below. However, as

---

[12] M. R. Kelly and H. Brooks: "From breakthrough to follow-through", in *Issues in Science and Technology* (Washington, DC, National Academy of Sciences), Spring 1989, pp. 42-47.

[13] R. Schultz-Wild et al.: *An der Schwelle zu CIM* (Cologne, RKW-Verlag, Verlag TUV/Rheinland, 1989), pp. 233-237.

regards economic incentives it is clear that the trend prevailing since the 1980s of lowering the relative price of labour — through deregulation, "give-back" collective agreements (whereby workers surrender concessions already acquired), *de facto* currency devaluation, subcontracting to low-wage countries and other measures adopted in some industrialised countries, particularly the United States — reduced the pressure to rationalise. There was less urgency to install capital-intensive equipment and to increase labour productivity.[14] A drawback is that taking the short-term view and neglecting investment in advanced technology can lead to a downward spiral. The use of advanced technology requires experience, skill and constant learning. It has been shown that enterprises with previous experience in automation are likely to have fewer problems with CIM technologies because their workforces tend to be familiar with the principles of automation and are thus better able to cope with advanced equipment. There is something like a virtuous circle in that innovation begets more innovation. The transition to CIM is therefore easier in countries with a well-educated and qualified manufacturing workforce.

If CIM is to have a wider impact, it will have to spread to medium-sized and smaller enterprises which still tend to shy away from the required high investments and generally lack know-how, but employ most workers in manufacturing. There are some signs that this may actually be happening, particularly in smaller enterprises working under subcontracting arrangements as component suppliers for larger enterprises in the automobile sector. Sometimes under pressure to conform to standards, delivery times and frequent component changes imposed by their major clients, sometimes on their own initiative to stay ahead in the technology race, such subcontractors have adopted CIM components such as CAD, CAQ, CNC machines, and so on.

---

[14] F. Bar et al.: "The evolution and growth potential of electronics-based technologies", in *STI Review* (Paris, OECD), No. 5, 1989, p. 46.

The socio-economic environment largely determines the approach chosen to implement CIM technologies, and the outcome. In market economy countries enterprises resort to them in order to survive in the face of competition. In the centrally planned economies of Eastern Europe management does not have the same incentive nor does it feel the same pressure to rationalise and innovate. However, planning authorities may decide that it is necessary to modernise industry, improve performance and raise labour productivity through CIM technology, and may allocate research and investment funds for that purpose. Such efforts have, for instance, resulted in a good number of pilot FMS and CAD/CAM installations in these countries which, from a technical point of view, are comparable with achievements in market economy countries. Nevertheless, there are high barriers to the diffusion of such advances throughout industry.

Conditions in the centrally planned economies vary. Those with a well-established machine tool industry, such as Czechoslovakia, the German Democratic Republic and the USSR, have made most headway in this field. Moreover, the economic reform and restructuring launched in the past few years aims to overcome the rigidities of central planning that puts emphasis on quantitative output and fulfilment of production targets. If some play of market forces is allowed, enterprises will be obliged to become more quality and cost conscious, and responsive to consumer needs. It is too early to say how far this recognition of existing problems, and the consequent policy reversals, will be crowned with success.

So far, however, the wide adoption and effective application of CIM technologies in these countries has been inhibited by the risk aversion and conservatism of industrial management, low labour costs, inappropriate incentive systems, resistance on the shop-floor, and weak linkages between component suppliers and equipment producers and between equipment suppliers and users. Moreover, in the allocation of resources priority has generally been given to the production of military hardware. Flexibility in the manufacture of capital and consumer goods was not demanded in the "seller's markets" of the centrally planned economies. These obstacles persist,

Computer-integrated manufacturing

aggravated by a shortage of investment funds and bottle-necks in materials supply. Advanced technology alone is clearly not sufficient for economic and industrial renewal.[15]

Figure 1. World-wide revenue from computers and related services (US$billion)

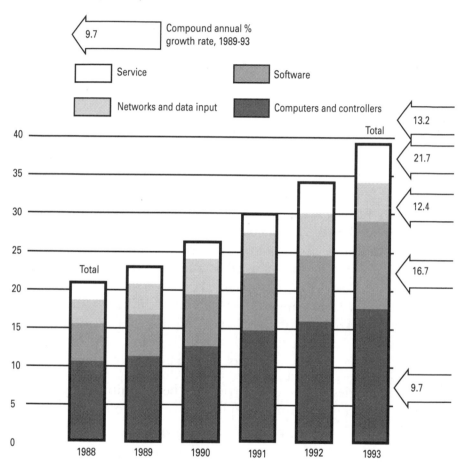

Source: Dataquest; cited in *Financial Times* (London).

[15] J. Baranson: *National differences in the business environment for automated manufacturing* (Geneva, United Nations Economic Commission for Europe, doc. Eng.AUT/SEM.8/R.61, 4 Sep. 1989).

The socio-economic perspective

17

What will the future hold? From 1983 to 1990 the world market for CIM equipment grew from US$15 billion to US$85 billion, with an annual increase varying between 20 and 25 per cent.[16] A forecast for the sales of computer hardware and software for CIM installations (excluding machinery, industrial robots, etc.) is given in figure 1. It looks as though the expansion will continue at a steady pace.

Industrial policy options

While the diffusion of CIM is mainly market driven, it depends a good deal on government policies in industry, education and training, finance, commerce, defence and telecommunications procurement, aerospace and international trade. Governments usually monitor technological developments and may be involved more or less directly in technology promotion and research and development depending on the social and economic system. At any rate, they set the framework and make political, economic and strategic choices that will influence the direction and speed of technological development. They can counteract bottle-necks and constraints, promote a favourable climate for technology diffusion and transfer, encourage investment, support the training infrastructure, and provide direct assistance through credit schemes, tax concessions, advisory services, research grants or financing of pilot projects. Governments may also foster links between research institutions and industry as well as attracting venture capital to start up enterprises in new technology fields.[17]

The difference in performance in CIM technologies between the United States and Japan has been partly traced to distinct

---

[16] M. Humbert: *Global study on world electronics* (Vienna, UNIDO, doc. ID/WG.478/2 (SPEC.)), p. 220.

[17] For a fuller discussion of such policy issues see J. Bessant: *Microelectronics and change at work* (Geneva, ILO, 1989), pp. 61 ff.

Computer-integrated manufacturing

government policies. In the United States defence procurement interests channelled research grants mainly into the development of highly sophisticated information technology and manufacturing processes for the production of complex military and space hardware. This specific technology has relatively limited civilian application. However, the United States has remained the leader in large advanced machinery and materials and in more sophisticated information technology applications. Japanese industrial policy concentrated on commercial applications of information technology and less-specialised NC and CNC machinery production. Such machines, which are in the lower price range, are within the reach of small and medium-sized enterprises. Incentives were provided to firms for installing this equipment. As a result, Japan has become the market leader in these fields and is reaping considerable benefits.[18]

In order to preserve or enhance the competitive position of the industries of their countries in world markets, governments have usually opted for active promotion of promising new technologies, and this includes CIM. A wide variety of measures and schemes addressing particular problems encountered in different societies are being applied. Most are designed to enhance the technological capability of industry and to develop the required human resources. This is not the place to review such measures or to assess their efficacy. Here it is sufficient to note that government action has become an often decisive support in the world-wide technology race, whose outcome will affect the future prosperity of countries and the quality of life of their people in no small measure.

CIM – An opportunity for developing countries?

While enterprises in industrialised countries struggling with the implementation of the CIM concept are faced with a mixed bag of failures and relative successes, developing countries are

[18] Bar et al., op. cit.

concerned about the widening technology gap between rich and poor nations. Would a transfer of CIM technologies help to close this gap, and would it make any sense? The answer depends on numerous factors which have different weightings in varying circumstances. There can be no hard and fast rule.

Frequently the example is cited of the newly industrialising countries (NICs) or threshold countries such as Brazil, India, the Republic of Korea, Malaysia, Mexico and Singapore, where some diffusion of such relatively mature technologies as CNC, CAD and robotics is taking place. Governments usually favour or support such investments. It is argued that these countries are benefiting from the skill-saving nature of these technologies and that there appears to be a much wider potential for their use, particularly as regards export-oriented industries.[19]

But is such a course advisable, realistic and practicable for the great majority of developing countries? Obviously, this depends primarily on the resource constraints and the pattern of factor endowment that these countries face, among which a chronic capital shortage and a cheap, abundant but low-skilled labour supply are the most outstanding. Moreover, the infrastructural supportive of CIM (e.g. adequate supplies of energy, raw materials, components and spare parts, reliable transport systems, easy access to high-technology markets, competent engineering consultancy services, research and training institutions) is lacking. Adequate maintenance of hardware and software can rarely be guaranteed. In addition, there is a need to ensure an adequate production volume and a large enough market to amortise the high investment needed for CIM. All this implies that there is generally not sufficient demand for sustained development in the field of high technology.

These are formidable obstacles to the introduction of advanced manufacturing technologies. At any rate, whatever industrialisation

[19] Edquist and Jacobsson, op. cit., pp. 205 ff.

goals are pursued, there can be no justification for squandering scarce capital on costly CIM projects destined to become white elephants. Turnkey operations proposed by eager vendors are the most likely to suffer that fate: they tend to function badly and become rapidly obsolete. It should be borne in mind that the investment required in complex CIM projects could easily be in the range of US$400,000 to US$600,000 per workplace.

Thus, entry barriers to CIM are high and there are many pitfalls. However, developing countries are confronted with the predicament that in many industrial activities, and particularly in the capital goods sector, their competitive advantage of cheap labour is declining. The latter is no substitute for the productivity and quality improvements achieved by highly skilled workers operating flexible automation systems, nor for faster delivery and service and the rate of product innovation achieved with advanced technology. If developing countries are to gain a foothold in competitive international markets for more complex products and do not want to be completely outdistanced, they will need to develop some local capabilities in this field. This entails difficult choices.

The most promising approach would appear to be to start with the relatively mature components of the CIM technology, i.e. CNC and CAD systems and microprocessor monitoring systems, but without going right away into the complexities of networking. Personal computer (PC) based systems for keeping maintenance records and running maintenance programmes may also bring great benefits. These would have to be incorporated in the manufacture of products for which a market could be found or developed, possibly in co-operation with enterprises from industrialised countries; and would have to fit into a strategic framework allowing a step-by-step adoption of more advanced technologies as experience is accumulated and learning proceeds. The adaptation of hardware and software to local conditions must be assured. Haphazard wholesale transfer of advanced technology usually spells failure.

Certain organisational innovations induced by CIM can be implemented without much costly hardware and software. They concern essentially the logical and efficient organisation of the production and information flow, changes in managerial responsibility and training of the workforce. They are discussed below in some detail. It has been proved that the productivity increases to be obtained through such methods can be spectacular.[20]

In 1989 the United Nations Industrial Development Organisation (UNIDO) launched a project on planning and programming the diffusion of industrial automation technologies in the capital goods industries of developing countries with a view to helping them formulate policies to assimilate such technologies. It is hoped that this project will provide a better and more realistic insight into the possible scope of the application of advanced technologies which, after all, have been conceived in the societal context of the highly industrialised countries. And even here they have often failed to reach their objectives. They can hardly be grafted on to industries without adequate preparation.

The impact on employment

There is encouraging evidence that by and large the aggregate level of employment in industrial societies is not greatly affected by the introduction of new technologies. The long-term trend of declining manufacturing employment observed in industrialised countries is certainly continuing, owing partly to technological innovations that eliminate unskilled work. It has, for instance, been found that FMS forming part of CIM saves direct labour in the range of 50-75 per cent when compared with conventional systems.[21] On the whole, however, job displacement and redeployment of workers in the course of innovation and rationalisation appear to balance each other, and where technological change goes along with strong

[20] J. Bessant and H. Rush: *Integrated Manufacturing*, Technology Trends Series No. 8 (Vienna, UNIDO, doc. IPCT.70, 1988), pp. 65 ff.

[21] IIASA, CIM study, op. cit.

economic growth, and expansion of markets and investment, it even tends to induce positive employment effects through the revitalisation of the economy. Japan's technology drive and growth pattern is a case in point. However, it would certainly be vain to pin exaggerated hopes on the introduction of CIM and other high technology and their spin-offs as an employment creation device. "Reindustrialisation through high technology" is a misleading concept.[22] Only a very small proportion of the labour force of the highly industrialised countries – some 2-5 per cent – are engaged in this advanced sector, and if past trends and experience are any guide, the proportion will rise only slowly, if at all. Although large enterprises can fairly easily muster the human and material resources required for CIM, the rather slow diffusion of advanced technology, particularly to small and medium-sized enterprises which in industrial societies employ the great majority of workers, prevents the effects from being more far-reaching. It was found, for instance, that in 1987 in the United States three out of four machine operators' jobs had not been directly affected by computerised automation.[23]

In this context it is also instructive to look at the distribution of the labour force by level of automation/mechanisation in the Federal Republic of Germany (figure 2). In 1985-86 only about 7 per cent of the labour force, i.e. 1.5 million employees, used programmable means of production such as data-processing equipment, PCs, NC/CNC machines, industrial robots and computerised medical equipment, while over 50 per cent continued to use simple tools. Since the previous survey was conducted in 1979, only a slight increase in the proportion of employees using programmable equipment had been observed. Advanced technology was obviously affecting the great majority of workers only gradually.

[22] Wobbe, op. cit.; K.-H. Ebel and E. Ulrich: *The computer in design and manufacturing – Servant or master? Social and labour effects of computer-aided design/computer-aided manufacturing* (Geneva, ILO, 1987), Sectoral Activities Programme working paper, p. 19 ff.

[23] M. R. Kelley and H. Brooks: *The state of computerized automation in US manufacturing* (Oct. 1988; mimeographed).

Figure 2. Use of means of production by employees in the Federal
Republic of Germany in 1985-86, by level of
automation/mechanisation

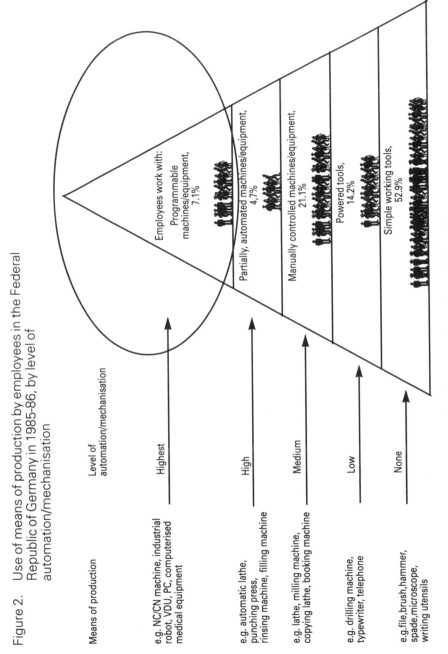

Means of production

Level of
automation/mechanisation

Employees work with:

Highest

e.g. NC/CN machine, industrial
robot, VDU, PC, computerised
medical equipment

Programmable
machines/equipment,
7.1%

High

e.g. automatic lathe,
punching press,
rinsing machine, filling machine

Partially, automated machines/equipment,
4,7%

Medium

e.g. lathe, milling machine,
copying lathe, booking machine

Manually controlled machines/equipment,
21.1%

Low

e.g. drilling machine,
typewriter, telephone

Powered tools,
14.2%

None

e.g. file, brush, hammer,
spade, microscope,
writing utensils

Simple working tools,
52.9%

Source: *Mitteilungen aus der Arbeitsmarkt- und Berufsforschung* (Nuremberg), No. 1, 1988, p. 28.

Computer-integrated manufacturing

It is clear that most manufacturing workers will not experience radical change as a result of CIM systems in the near future. Nevertheless, it must be expected that CIM will accentuate the already existing labour market segmentation. A core of highly skilled workers and qualified technicians, engineers and professional workers manage and operate such systems and are increasingly indispensable. They are generally well paid and their working conditions are stable thanks to their position as knowledgeable workers and experts, but access to this core is more and more difficult to achieve. Ancillary workers have little opportunity for upward mobility unless they possess or acquire the necessary skills, and their conditions of work are also more precarious. This situation may eventually lead to social conflicts.

Finally, the labour market situation itself may reinforce the trend towards CIM. It has been observed that labour shortages are often a significant driving force of automation. There may be a lack of skilled workers, technicians and engineers or simply reluctance on the part of a better-educated workforce to perform menial, tedious and repetitive tasks in manufacturing. This acts as a strong incentive for enterprises to make industrial work attractive, to introduce automation and to change work practices and organisation. There is much evidence in support of this view from countries such as the Federal Republic of Germany, Japan, Sweden and the United States.

THE TECHNICAL PERSPECTIVE

The state of the art

Computer-integrated manufacturing is many things to many people. Enterprises have to seek their own solutions to the multitude of technical problems in response to their specific requirements. Depending on the complexities of the manufacturing process and the existing organisational pattern, these problems may well be formidable. Existing model solutions are not generally applicable and only show that integration is feasible to a large extent and under specific circumstances — though often at a high price. The final configuration of CIM depends on such factors as production volume; product variations; product life cycles; product size, weight and complexity; number of parts; size of plant; available human resources; and relationship with suppliers and customers.

It is, therefore, useful to keep in mind what can be gleaned about the state of the art from various studies of the introduction of CIM.

There are first of all a number of technical shortcomings. A great deal of research and development goes into eliminating technical bottle-necks which hinder a wider application of CIM. It appears to be a general phenomenon that the potential for flexibility of existing flexible automation systems is rarely fully used. Considerable progress is being made on a broad front and, *a priori*, one cannot say that the remaining problems are insoluble. However, there is still many a hard nut to crack and it is not sure when, and at what cost, effective solutions will be found.

The technical rationale of CIM consists of linking, through networks, a wide variety of numerically controlled machine tools and industrial robots, process control computers, automatic transport and storage facilities, quality assurance, design and production planning functions and management information systems. So far, the different subsystems of CIM such as CAD, CAM, CAP, and so on, have been treated as isolated functions with their own requirements, logic and software — resulting in "islands of automation" in factories. Although there are many partial solutions to bridging these functions, few truly integrated large systems, comprising all functions of designing, manufacturing and assembling products with a great number of components, as yet exist.

When looking at integrated manufacturing, it is helpful to keep in mind that there are, in fact, successive levels of integration which are progressively difficult to achieve. Figure 3 shows this progression. It also makes clear that in practice an incremental approach to the implementation of CIM has the best chances of success because it enables the workforce to assimilate the specific techniques gradually, while building the system from the bottom to the top.

Figure 4 visualises the convergence of technological trends towards CIM. The actual combination of all technology components into CIM is forecast for 1990, which may well prove a rather optimistic assumption.

Software problems

Information technology, involving a large amount of storage, retrieval, processing and communication, lies at the heart of CIM, as shown in the following example:

... a machine controller takes in information (communication) about the state of the process it is controlling. It then compares this with other information in its memory about the desired state (storage/retrieval) and calculates (information processing) the necessary corrective action. Finally, it sends information (communication) back to the process to bring it back into line.[1]

Other CIM functions work along the same lines using digital representation of information. A great deal of complicated software is needed to make computers execute the wide variety of functions required.

Networking of systems is generally in its initial stages. The development of computer software for such tasks is painstaking and the widespread adoption of standards, which are vital for linking computer hardware and automation equipment of various makes, is lagging behind. Constant complaints are made about the proliferation of proprietary operating systems, communication options and control programmes. The existing chaos holds back networking as manufacturers hesitate to commit themselves; it also increases costs.

Vendors and users of flexible automation equipment are painfully aware of this situation. General Motors, one of the largest clients, has therefore pioneered an industry-wide standard — the manufacturing automation protocol (MAP) for the linking of stock control, robotics, CNC machinery and quality control — and will buy only equipment conforming to these standards. The company has agreed with the aircraft manufacturer, Boeing, which uses the technical office protocol (TOP), to make the two systems compatible (TOP is largely complementary to MAP). These plans have aroused some controversy. Moreover, the systems have been criticised for being too slow and exclusive, and for being tinged with the image of company proprietary standards rather than tying in with industrial robotics. Some companies are committed to other communications standards which are not fully compatible with MAP or TOP. In fact, Digital Equipment is the *de facto* market leader with its DECnet

[1] Bessant and Rush, op. cit., p. 6.

Figure 3.   Levels of automation

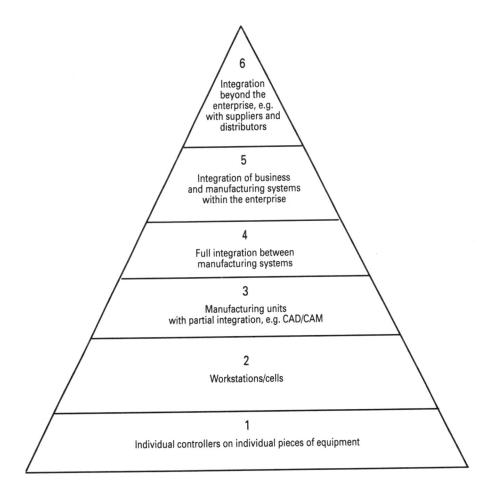

Source: J. Bessant and H. Rush: *Integrated Manufacturing*, UNIDO Technology Trend Series No. 8 (Vienna, UNIDO, 13 Oct. 1988), p. 8.

Computer-integrated manufacturing

Figure 4. Convergence of technological trends towards CIM

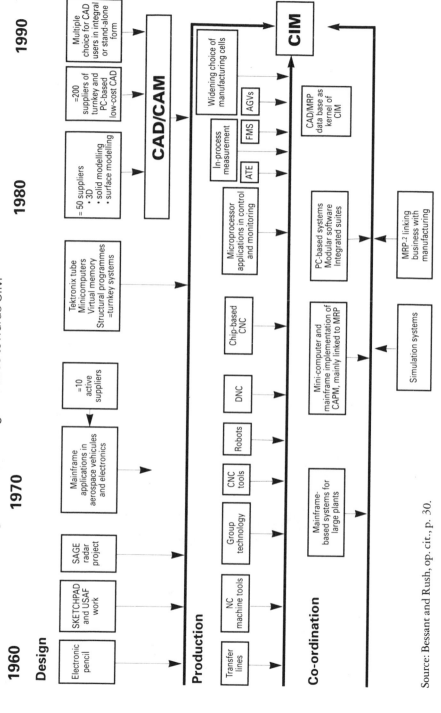

Source: Bessant and Rush, op. cit., p. 30.

plant-networking system, itself based on Ethernet. Manufacturers that have heavily invested in such proprietary systems are, of course, hesitant to switch to another. Nevertheless, it should be noted that MAP is making steady advances and is now being used in several industries, as shown in figure 5.

The International Standards Organization (ISO) has set itself a more ambitious objective; it is promoting the Open-Systems Interconnection (OSI), which is to enable all computer systems to communicate with each other. However, OSI also encounters delays and acceptance problems.[2] It appears that a great deal of research is

Figure 5.   MAP networks installed by industry in 1989

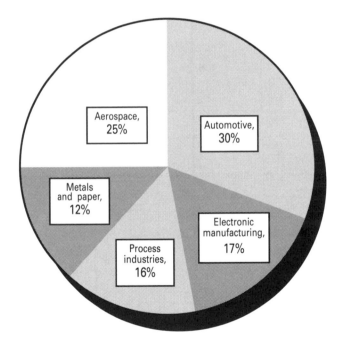

Source: J. Trotta and M. Greaves: "The future of MAP", in *Automation* (Cleveland, Ohio), July 1989, pp. 34-35.

[2] A. Cane: "Computers in manufacturing", special survey in *Financial Times* (London), 2 June 1987, pp. I and II.

still needed to arrive at satisfactory world-wide standards, even though a significant amount has already been undertaken (for example, within the European ESPRIT programme).

Compatibility of equipment and systems is certainly a crucial problem, but there are others. Much information required in the production process is not suitable for coding, computer processing and transmission. This means that available data are not always complete or reliable. However, full integration of the production process means that it must be predictable. Stringent procedures that have to be formalised are required; these cover process planning, tool supply, production planning and logistics. Such procedures are, of course, the antithesis of flexibility, but they are necessary because robots or process controllers can neither think nor anticipate.

CIM needs to be custom-built by the enterprise itself. Therefore, the largest part of CIM software is produced in house by users. This affects its quality. There is, indeed, a high risk of hidden errors that may create disturbances and malfunctions in the most unexpected manner. This is aggravated by the fact that even careful analysis and operational testing for reliability and safety will not necessarily show up inherent failure modes, which mainly concern rare events.

Further problem areas are unreliable software and automation equipment, rudimentary sensory abilities of industrial robots, and poor data quality and accessibility. Data inconsistency makes the establishment of common data bases and the sharing of data difficult. The need for constant updating of data bases is another drawback. The result of such combined difficulties is that systems are vulnerable and break down frequently, often for more than one-third of the available time.

The difficulties of achieving perfect process control through computers is well illustrated by the following comment concerning the machining of metals:

It is possible to construct a mathematical model of the machining process, which can be used, given the availability of certain data, to calculate the optimum way of cutting metal. This is the theoretical picture. In practice, however, this mathematical model is unlikely to cope with all but the simplest and most carefully defined situations. The reality is that the data used in such mathematical models are subject to large degrees of uncertainty, the cutting process itself is not fully understood and cannot therefore be accurately modelled, and the work is subject to a large number of unpredictable disturbances.[3]

Moreover, the integration of process information subsystems, which is in fact the corner-stone of CIM, remains a tricky undertaking despite the advances in automation protocols already mentioned. An example of such a process information system is given in figure 6.

In theory, such a system will support the total control of a plant. In practice, the numerous functions to be integrated by software result in considerable complexity and even small technical failures or operator errors can escalate in unexpected ways and cause the whole system to collapse. Thus the safety risks of even small errors can be considerable. A study of CIM carried out by the International Institute for Applied Systems Analysis (IIASA) has assembled empirical data and arrives at the following conclusions:

The major sources of failures were found in the interfacing software of different subsystems of an integrated system, in the software for managing rare events as well as in operator errors, which escalated in an unexpected way. More than one-third of them can be traced back to design and design errors, and one-fourth to unsatisfactory training and poor instructions and guidelines. . . . Totally, the increased use of information technologies seems to improve safety and decrease unintended impacts. However, there is an increasing risk of rare shock events (low probability-high impact events) with major shock impacts, due to the possibility of serious, escalating failures of software systems. . . . In manufacturing . . . they may lead to economic catastrophes, accidents and risks for occupational safety.[4]

It might be added that optimal CIM requires complete real-time data processing. However, current CAM systems process about 85-95 per cent of data by batches, which is considered

---

[3] P. T. Kidd and J. M. Corbett: "Towards the joint social and technical design of advanced manufacturing systems", in *International Journal of Industrial Ergonomics* (Amsterdam), No. 2, 1988, p. 307.

[4] J. Ranta: *Tomorrow's industries, structural changes and the need for work organisations* (Laxenburg, IIASA CIM Project, 1989), unpublished paper, pp. 17-18.

Figure 6.   An integrated process information system

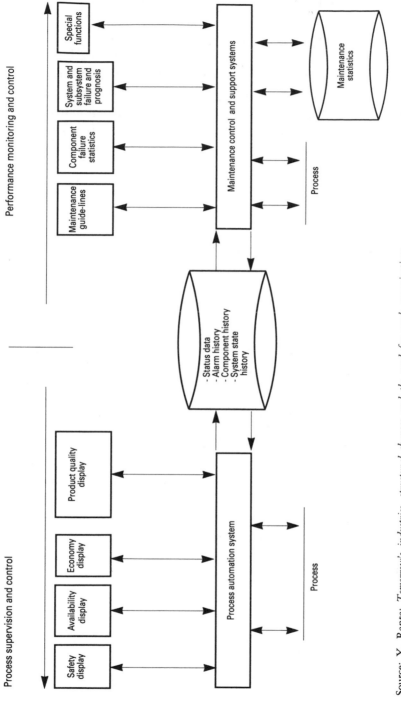

Source: Y. Ranta: *Tomorrow's industries, structural changes and the need for work organisations*, International Institute for Applied Systems Analysis (IIASA), CIM project, unpublished paper.

insufficient.[5] CIM is praised for its flexibility. This is justifiable to a certain extent, in particular when it comes to machining families of parts. However, CIM is inflexible with respect to alterations in batches and when process innovation is contemplated. In fact, every change in a customer's order or equipment, tools or materials has first to be modelled in the computer system. Much depends, of course, on the type of system chosen. By now, many enterprises have found that they have acquired an integrated system that they cannot adjust to changing needs. This tends to halt continuous process improvement.[6]

Further system improvements

A number of technical remedies are on the way. In fact, science and technology move faster than industry's capacity to apply new knowledge. The development of fault-tolerant computers and adaptive control systems is making progress. The sensory abilities of robots are constantly improving. Greater density of data storage and faster processing speed of computers will help to build more "intelligent" systems. In particular, the development and use of more highly performing workstations as an engineering tool will facilitate the implementation of CIM. They should give their users access to all relevant data bases and enable them to intervene in all phases of the production process, and assist in decision-making and in data transformation. Not only are they invaluable for organising the interaction of design and manufacturing, which is a notoriously difficult proposition, but they are expected to become the main

[5] G. G. Axelsson: *Computer-integrated manufacturing: How to get there*, International Management Institute, Technology Management Unit, Report No. 002 (Geneva, Oct. 1985), p. 48.
[6] J. Cole, T. Friscia and J. Heaton: "Sorting through the confusion and chaos of systems integration", in *Automation* (Cleveland, Ohio), Oct. 1989, pp. 48-56.

instrument for integrating all operations and to facilitate team-work.[7] Much research is also being undertaken with the objective of reducing the escalating cost of software development through the design of software engineering tools and techniques.

Moreover, some promising developments in "artificial intelligence" and expert systems may well help to make systems more tolerant and responsive to faults. Expert systems, already being used in standardised procedures and routine checking, are beginning to play a role in maintenance, fault diagnostics, production control, quality assurance, planning, scheduling, design (notably CAD), and support software development and training. Nevertheless, such systems have their limitations. While they embody knowledge extracted from experts and apply reason according to given rules, there are limitations to formalising human abilities and sensorial experiences, not to mention open and complex industrial processes. Furthermore, common-sense knowledge is not programmable, nor do expert systems have intuition or practise associative thinking. They can therefore support, but not replace, human decision-making.

The overestimation of the performance capacity of expert systems, which tend to reduce reality to the possibilities of a system and which make workers dependent on often not verifiable knowledge and data, thus depriving them of responsibility, may well dampen the enthusiasm for such systems in the future.[8]

The Informatics Society (*Gesellschaft für Informatik*) of the Federal Republic of Germany warns:

Expert systems constitute a new programming paradigm to which an almost unlimited capacity for solving complex problems is imputed *a priori*. Even if the formulation of poorly defined problem areas does indeed frequently prove more successful using facts and rules and heuristic methods for problem-solving than with conventional programming methods, these systems, too, can only assume a

---

[7] G. Schaffer: "Workstations: Windows into CIM", in *American Machinist* (New York), Mar. 1989, pp. 75-83.

[8] U. Hillenkamp: *Expert systems — Present state and future trends: Impact on employment, working life and qualifications of skilled workers and clerks*, ILO/Federal Republic of Germany Project on Expert Systems and Qualification Changes (INT/86/M03/FRG) ILO, Management Development Branch, Geneva, 1989, p. 116.

supportive function. To hope that they will render superfluous the responsible evaluation of results by human beings is a dangerous fallacy, since expert systems are specifically designed to allow uncertain results without these always being recognisable as such.[9]

Another way of minimising technical problems is the initial design of products in such a way that they can be processed and assembled with the flexible automated equipment available. This means simplifying and reducing the number of components, streamlining assembly operations and generally taking into account the ability of machines and robots at the design stage.

As a wide range of system improvements comes on to the market or is designed by manufacturers' technical offices for their own use, a suitable and informed choice among many technological alternatives becomes more and more important. This must take account of the strategic objectives of the enterprise and must respond to criteria deemed to be effective from an overall managerial and organisational perspective. Managing the introduction of system innovations is, therefore, a key managerial function. This has many implications which deserve closer scrutiny.

[9] International Federation of Automatic Control (IFAC); Committee on Social Effects of Automation: "Computers and responsibility: Result of Working Group 8.3.3 of the German Gesellschaft für Informatik on 'Responsible Use of Information Technology'", in INFO Pack No. 5, Aug. 1989, p. 3.

THE MANAGERIAL AND ORGANISATIONAL PERSPECTIVE

CIM – The end of chaos or its beginning?

M anagement attitudes on how to cope with CIM vary. In part they mirror the national idiosyncracies and different industrial backgrounds evoked above. Thus the technocentric approach results in management strategies that neglect or underrate the human factor in production. This tendency is frequently aggravated by short-term profit considerations which are one of the major stumbling blocks to sound technology planning and management. The introduction of CIM requires long-term strategic thinking. From the managerial point of view it is essentially an organisational quandary: how to create order out of potential chaos. Equipment needs to be carefully selected and compatibility ensured. However, the fundamental problem is how to reshape existing production processes, alter organisational boundaries and make them permeable. This requires redesigning the information and data flow to foster decentralised decision-making. The difficulties of actually accomplishing this in existing organisations, to make them more effective and efficient, should not be underestimated.

It must not be forgotten that the relatively slow productivity growth in manufacturing observed in recent years has much to do with haphazard computerisation and system incompatibilities within organisations. A down-to-earth comment reported by Bessant puts it well: "When you put a computer into a chaotic factory the only thing you get is computerized chaos!"[1]

At the same time, the introduction of CIM may well act as an antidote to poor management practices. This is where the human factor comes in. Industrial case studies and experience accumulated so far clearly indicate that a pragmatic management approach which advances step by step, builds up the skill, responsibility and motivation of the workforce, invests in people operating the systems and relies on the human factor to make it flexible, has consistently paid off best.

This implies going beyond Taylor's vision of "a system in which the workman is told in minute detail just what he is to do and how he is to do it, with any improvements which he makes upon the orders given to him being seen as fatal to success".[2] The principal characteristics of non-Taylorist as opposed to Taylorist systems are given in table 1.

The conviction that this is really the best course seems to be lacking in many management circles; otherwise a more systematic and consistent effort would be made to enhance the human factor. At any rate, it has been found that in general CIM is not introduced primarily to "humanise" work. The motives and expectations of management have to do mostly with stock reduction, greater transparency of the organisation, reduction of lead time, closer adherence to deadlines, saving of personnel, greater marketing flexibility, increased capacity use, higher product quality or the need to keep up with technological developments — all leading to cost reduction and higher productivity.

[1] Bessant and Rush, op. cit., p. 35.

[2] F. W. Taylor: "On the art of cutting metals", in *Transactions of the American Society of Mechanical Engineers*, No. 28, 1907, pp. 31-350.

Computer-integrated manufacturing

Table 1. Characteristics of Taylorist and non-Taylorist systems

Taylorist systems	Non-Taylorist systems
Centralisation of control and decision-making	Decentralisation of control and decision-making
Operator told how work is to be carried out	Operator decides how work is carried out
Standardisation of working methods	Operator develops own working methods
One best way of doing work	Many good ways of doing work
Functional specialisation	Multi-skilled operator
Simplification of work	Variety of complex tasks
Imposing methods and solutions on people	Consultation and participation

Source: P. T. Kidd and J. M. Corbett: "Towards the joint social and technical design of advanced manufacturing systems", in *International Journal of Industrial Ergonomics* (Amsterdam), No. 2, 1988, p. 309.

An expert survey conducted in the Federal Republic of Germany in 1987 is unequivocal on this point, as shown in figures 7 and 8.[3]

Improved working conditions thus tend to be an incidental by-product of CIM, and to assume a very low priority. In fact, working conditions may even be neglected or grow worse, particularly where automated machinery is used to enforce an accelerated pace of work, where only residual tasks are entrusted to workers or where the new tasks created by computerisation involve a high degree of stress. In this context it is significant that in the United States there is evidence of a considerable increase in repetitive strain injury attributed to poor job design. It is, for instance, alleged that about half the members of the United Autoworkers Union, i.e. some 500,000 workers, suffer from such health disorders. Keyboard operators tend to be affected more than other occupations.[4] This may

[3] E. Köhl, U. Esser and A. Kemmner: *CIM zwischen Anspruch und Wirklichkeit — Erfahrungen, Trends, Perspektiven* (Cologne, TÜV-Verlag, 1989).

[4] "Experts see epidemic in repetitive motion", in *AFL-CIO News* (Washington, DC), 24 June 1989, p. 3.

Figure 7. Frequency of basic enterprise objectives (percentages) (each expert specified five objectives)

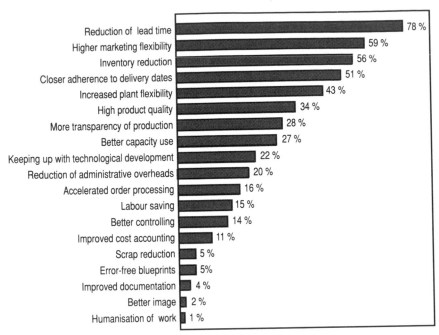

Reduction of lead time	78 %
Higher marketing flexibility	59 %
Inventory reduction	56 %
Closer adherence to delivery dates	51 %
Increased plant flexibility	43 %
High product quality	34 %
More transparency of production	28 %
Better capacity use	27 %
Keeping up with technological development	22 %
Reduction of administrative overheads	20 %
Accelerated order processing	16 %
Labour saving	15 %
Better controlling	14 %
Improved cost accounting	11 %
Scrap reduction	5 %
Error-free blueprints	5%
Improved documentation	4 %
Better image	2 %
Humanisation of work	1 %

Source: E. Köhl, U. Esser and A. Kemmner: *CIM zwischen Anspruch und Wirklichkeit – Erfahrungen, Trends, Perspektiven* (Cologne, TÜV-Verlag, 1989), p. 104.

well indicate that, instead of alleviating physical strain, much automation has led to increased work pacing of residual tasks by machines. If confirmed by further research, this would certainly constitute a rather perverse result of technical advance.

A new management style

A management style that allows production personnel greater autonomy and restores initiative to the shop-floor may well mean a break with entrenched principles and thus be seen as a threat to vested interests and the power structure in an organisation.

Computer-integrated manufacturing

Figure 8.   Influence of CIM technology on reaching objectives
(percentages)

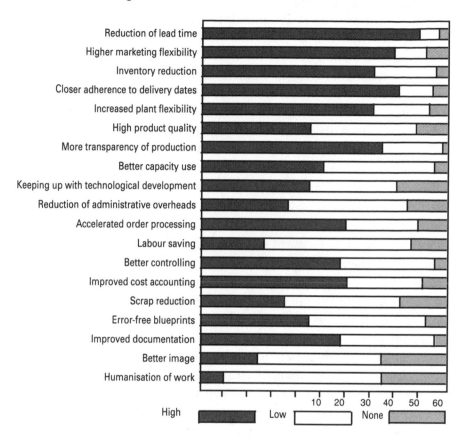

Source: Köhl, Esser and Kemmner, op. cit., p. 123.

Although it is not surprising that as a rule everything is done to avoid such clashes, it is, in fact, possible to switch to new technology without making fundamental organisational changes, and to keep established divisions and top-heavy hierarchies in place. Information technologies may also be used to institutionalise and even reinforce ineffective and counterproductive management practices such as excessive centralisation of decision-making or abusive monitoring of individuals. This is, of course, costly and leads to mediocre results while prolonging the life of organisational dinosaurs. It defeats the primary purpose of CIM, i.e. the integration of all functions, which requires "vertical and horizontal synchronisation of departments, people, machinery and processes in the flow of information and material".[5] In such a system the necessary flexibility can be achieved through decentralisation of information and responsibility within a given framework in order to achieve small and fast control loops. This enables the production system to respond rapidly to market demand, particularly in the case of multiple product options.

The effective introduction of CIM requires a clear strategy that has the backing of both senior management and the rank and file. Obviously nothing much can be done without the consistent support of top management; however, the stumbling-block may prove to be middle management, who often dread the destabilising effect of advanced technology more than do skilled workers and technicians since they stand to lose part of their influence when established hierarchies are dismantled and all needed information is available directly "on-line" to all participants in the production process. This emphasises the need for top-level technology management, a function frequently neglected because legal, financial and marketing aspects tend to dominate decision-making at the top. It is not enough to let middle management acquire new technology and then resort to crisis management at the top when bottle-necks occur in the organisation or the middle managers lack the necessary skills for handling the technology. It is, of course, not enough for top management to announce a CIM strategy; this can only be implemented at the lower levels of the hierarchy with the commitment of everybody involved. Such commitment must be assured through consultation.

The implementation of such a strategy is by no means easy and much depends on the specific national, cultural, industrial relations and labour market context in which enterprises operate. It will succeed only if the workforce is adequately prepared and willing to

[5] Warnecke, op. cit.

co-operate. Little will be gained from thrusting an unwanted autonomy and a participative structure ordained by top management on a reluctant workforce unable to perceive the advantages of new forms of organisation and to recognise its own interest in adopting a new approach. Indeed, many workers may find it convenient to rely on detailed instructions without assuming much responsibility and may refuse to accept multi-skilling. Taylorism counts on this attitude, which has no place in CIM systems. However, as it is widespread many efforts to introduce autonomous group work may be doomed to failure.

Some commentators have traced existing problems to the lack of managerial competence.[6] Managers' knowledge of advanced manufacturing systems is frequently limited, even when they have received a technical education or are professional engineers. Owing to the advances in manufacturing technology, and especially information technology, the professional knowledge and experience that managers have acquired rapidly become obsolete unless they are continually exposed to shop-floor experience and the latest techniques. Consequently, potential users of automation equipment, often fearing that they will not be able to muster and constantly update the know-how needed for operating the equipment, fall back on outside consultants and equipment suppliers. They naturally tread carefully in unknown territory and avoid incalculable risks.

Managers are also under pressure to justify the high capital expenditure required for implementing CIM. By the standards of a short-term return-on-investment (ROI) approach, the financial feasibility of most CIM projects is doubtful despite the hypothetical long-term economic advantages outlined above. In fact, there are no generally agreed methods for making reliable cost-benefit analyses of CIM, and the cost of full-scale CIM implementation is often considered to be prohibitive, especially when the cost of tailoring the system to the enterprise's specific uses is added to the cost of the

[6] R. Hodson and J. Hagan: "Skills and job commitment in high technology industries in the US", in New technology, work and employment (Oxford), pp. 113-124.

equipment. It is also feared that CIM will be inefficient to use and expensive to maintain because technical change constantly requires the replacement of parts of the system, a particularly difficult task in an integrated system. To this should be added that in present flexible manufacturing systems fixed costs constitute about 70 per cent of the total outlay. This is one indication of the high risks that management takes when installing CIM.

It must also be considered that the implementation of major organisational changes is not without costs. They are, in fact, substantial and are said to exceed the costs of building the system, i.e. the acquisition of hardware and development of the corresponding software, by two to three times; they also have a tendency to grow.[7]

There is, of course, the other side of the coin: risks are balanced by opportunities if the expected economic benefits of CIM materialise. Moreover, in the years ahead the capital outlay is bound to decrease as cheaper systems come on the market. This will make CIM technologies more attractive for small and medium-sized enterprises. It has also been estimated that CIM plants could break even at 30-35 per cent of capacity utilisation as against 65-70 per cent in the case of conventional plants. In addition, the planning of CIM and at least partial implementation could help management to improve the organisation of the production process and the flow of communication. A leaner management will be able to cope better with essential planning and control tasks and to devote more time to functional tasks such as product development and delivery and customer service. This will enhance productivity and speed up reaction to market demand.

The most promising prospect of all is probably that CIM technologies will enable management progressively to dissociate factory operation time from actual working time through a wide range of flexible working arrangements. This will considerably

---

[7] R. Roberts and A. Hickling: "Computer integrated management?", in *Multinational Business*, No. 2, 1989, pp. 18-25.

enhance the utilisation of fixed capital and increase the return on high investments, thus diminishing risks and making enterprises more competitive. It is also likely to ease recruitment problems, particularly among qualified women workers who will be able to choose work schedules to fit in with their family responsibilities.

Introducing CIM: A strategic decision

Costs are not the only element to be considered in making the strategic decision to introduce CIM. A responsible and forward-looking management may well conclude that the company cannot afford to be left behind in the technology race and that it must meet research and development expenditures for process technology and investment in CIM in order to remain competitive in the long run. Clearly, much depends on the specific situation of the enterprise. The first question to be answered is whether CIM can really solve existing manufacturing and marketing problems and safeguard the future of the enterprise.

If it is decided to implement CIM, management's essential task is to overcome hierarchical rigidities and organisational resistance to the change, from the shop-floor up through all layers of the organisation. To streamline an organisation and make it fit for CIM may be a considerable challenge but may be well worth the effort. The findings of a variety of surveys concur on this point; manufacturers who have introduced advanced manufacturing systems attribute between 40 and 70 per cent of the total improvement achieved to better logistics and organisational changes. In other words, the main benefit does not necessarily stem from the sophisticated and integrated technology itself but from the reform of management and production practices and from a more transparent and efficient organisation.[8] Such significant opportunities may well

[8] B. Haywood and J. Bessant: *The integration of production processes at firm level*, mimeographed, Brighton Polytechnic research paper (Brighton, Sussex, 1987).

justify a departure from the reassuring, tried and simple precepts of Taylor's "scientific management" – and the human-centred approach to CIM means nothing less.

The various stages involved in a CIM strategy are summarised in table 2.

In employers' circles there is, in effect, growing awareness that the changing world of work and the increasingly competitive business environment require new management approaches in manufacturing. The need to tap available human resources and to

Table 2. Stages of CIM strategy

Stage		Action
(1)	Define business strategy	Market analysis Competitor analysis Business planning
(2)	Carry out manufacturing resources audit	Review all current operations Cost distribution analysis Opportunity audit
(3)	Develop broad CIM strategy	Explore options (including organisational innovations) Priorities/sequence Planning
(4)	Identify CIM architecture	Specify equipment Specify communications Specify infrastructure
(5)	Project planning	Define tasks Allocate responsibilities Define links between projects
(6)	Implementation	Resources Timing Controls Organisational development
(7)	Continuing audit and monitoring	

Source: Bessant and Rush, op. cit., p. 65.

motivate the workforce is becoming a widely shared objective. This is exemplified by a set of guide-lines and recommendations issued by the Federation of Metal Trades Employers *(Gesamtmetall)* of the Federal Republic of Germany,[9] which stresses that people's attitudes to work are changing and that it is not enough for them simply to earn a living; they desire meaningful work. Employers should see this change in attitude as an opportunity to be seized and should seek the commitment of their workforce to a common goal. The comprehensive recommendations refer to the use of technology, work content and work organisation, further training, flexible working time, occupational safety and health, information and consultation, and modern management. It can only be hoped that such principles will be widely applied in practice.

The National Economic Development Office of the United Kingdom published guide-lines for employers in a concise form which essentially cover the same ground.[10] They are quoted here as an example:

Consult widely with employees at the earliest practicable stage	TO	remove anxiety and encourage participation and commitment
Obtain visible commitment from top-level management	TO	ensure objectives are set, resources committed and plans approved
State objectives and benefits clearly	TO	help users identify with business success and see the role of the new technology in achieving this
Emphasise that systems can reduce the drudgery in jobs	TO	gain user acceptance
Guarantee no compulsory redundancy	TO	remove anxiety

[9] Gesamtmetall: *Mensch und Arbeit* (Cologne, Edition Agrippa, 1989).

[10] National Economic Development Office (NEDO): *The introduction of advanced information systems* (London, NEDO/Her Majesty's Stationery Office, 1985).

Plan user involvement	TO	get the system right and extend user involvement
Structure the project so that users and systems specialists work in parallel	TO	ensure understanding grows at the same rate
Provide training about the technology	TO	raise the general level of awareness and increase the effectiveness of the user's contribution
Ensure the user manager understands the system	TO	enable him to support his staff
Give users early, "hands-on" training	TO	allay any fears about their own competence or that of the system
Users and suppliers of technology should monitor the system together	TO	identify recurrent problems and possible improvements
Prepare training for the right time	TO	take advantage of the long lead time while avoiding loss of effectiveness if training is given too early
Fill new jobs internally	TO	conserve existing knowledge and maintain motivation

Source: Köhl, Esser and Kemmner, op. cit., p. 123.

Skill requirements for CIM

The foregoing observations imply that there should be a rise in the level of skills of shop-floor workers or that different skills will be needed, despite the fact that some of their present skills will become obsolete and trends towards the division of labour will be reversed. CIM requires versatile craftsmen and technicians, computer and software experts, mechanical and communications engineers and, in general, people who understand production methods and the system and are capable of handling a great deal of technical information and of taking decisions on the spot. These requirements go far beyond simple machine-minding, since only qualified people can ensure maximum utilisation of the costly equipment. There is little room for unskilled workers such as assemblers, labourers, machine loaders and transport workers. CIM also renders redundant clerical workers engaged in ordering parts and materials and scheduling the workload of machines.

Specific skill requirements and training needs depend to a large extent on the organisational changes triggered by CIM and the particular CIM configuration and network strategy chosen by individual enterprises. There will be new working procedures, task structures and work content, as well as new relationships among different technical functions. It is not impossible that in some instances skill requirements will be reduced, when equipment components embodying information technology simplify or facilitate tasks, e.g. maintenance of electronics equipment by replacement of components such as electronic cards instead of difficult repairs on site. On the whole, however, trouble-shooting and maintenance work tend to become more complex because of the integration of electronic, electrical, mechanical, pneumatic and hydraulic functions.

Computer-aided design (CAD) is an important component of CIM. It speeds up the design and production of drawings and enables the performance of operations that would be impossible without a computer such as the rapid production of design variations, direct inclusion of calculations, simulations of functions, very complex designs (e.g. computer chips) and direct transmission of machining data. It also expedites routine work such as detail drawing, information searches, calculations (finite element analysis) and the establishment of lists of workpieces and control work. As regards the new skill requirements for designers and draughtsmen, a comparative ILO study concluded:

The main new skill requirements for staff working with CAD systems appear to be "computer literacy" and higher mathematical and analytical skills, especially a good understanding of the principles of analytical geometry and the application of co-ordinate systems, together with an open mind and a high degree of accuracy and attention to detail. . . . Traditional draughtsman skills nevertheless appear to retain their validity and importance — at least for the time being — since the main tasks remain unchanged. Because they are relieved of much routine work draughtsmen and designers should have more time for creative work, but in practice existing software and rules and macros stored in data banks may seriously limit their options.[11]

In the Federal Republic of Germany a detailed study of a large FMS points out the extent to which maintenance skills (especially the newer ones such as systems analysis and diagnostics) contribute to the utilisation of advanced manufacturing systems. An analysis of more than 6,000 hours of operation revealed that more than half the system down time was due to unscheduled stoppages or breakdowns. Of the time taken to repair and bring the system back into operation, around half was taken up in diagnosis. The conclusion was that:

> . . . the more complex and automated the systems were, the higher the skills level of the maintenance specialists had to be to achieve reasonable failure rates and implement facility improvements . . . and . . . the lower the personnel levels were (producing with automated facilities), the broader the educational background of these workers (operators and maintenance) had to be.[12]

Middle-level managerial jobs are also bound to decrease or diminish in CIM systems because of the general dissemination and free flow of information. There tend to be fewer hierarchical levels and demarcation lines, and fewer co-ordinating tasks. The emphasis is on planning, creativity, anticipating problems, less formal communications, teamwork and interaction, and much less on giving instructions. Excessive monitoring of workers (which is technically possible) is best avoided because it can antagonise the very people needed to run the systems. There is a new world for team leadership in CIM which requires from managers a subtle combination of human, conceptual and technical skills.[13]

[11] K.-H. Ebel and E. Ulrich: "Some workplace effects of CAD and CAM", in *International Labour Review* (Geneva, ILO), May-June 1987, p. 365.

[12] G. Handke: "Design and use of flexible manufacturing systems", in *Proceedings of the Second International Conference on Flexible Manufacturing Systems* (Kempston, Bedfordshire, IFS Publications, 1982).

[13] O. L. Crocker and R. Guelker: "The effects of robotics on the workplace", in *Personnel* (New York, American Management Association), Sep. 1988, pp. 26-36.

The implementation of CIM requires continuous learning and alertness at all levels. Since CIM is essentially based on microelectronics and information technology, it therefore demands much theoretical understanding, abstract thinking and comprehension of work methods. The development of cognitive skills, as well as analytical, problem-solving and logical capabilities, has a high priority; and such qualifications are in scarce supply anywhere. In addition, organisational, economic, social and communication skills are called for as an increasing proportion of resource management is carried out at the shop-floor level.

While higher skill requirements and greater versatility — and finally more intellectually demanding and interesting jobs — will be the rule in CIM installations, particularly for the core workers indispensable to keep systems operating, it is also true that the skill content of jobs depends largely on the particular category of enterprise, the production process and the work organisation chosen. For some workers concerned it might appear to be quite a different proposition. Certain optimistic views and generalisations about multi-skilling, job enlargement and job rotation need to be taken with a pinch of salt. When certain "skills" can be acquired in just a few days, multi-skilling may turn out to be a euphemism for work intensification. To quote an automobile worker: "The jobs are just the same as before, you just do more of them".[14]

New types of work organisation

Computer-integrated manufacturing enables management to integrate functions and tasks and to create new types of jobs. It makes job enlargement and enrichment a realistic possibility through broadened work content, better information and more

[14] P. J. Turnbull: "The limits to 'Japanisation' — Just-in-time labour relations and the UK automotive industry", in *New Technology, Work and Employment* (New York, American Management Association), Vol. 3, No. 1, 1988, pp. 7-20.

decision-making power. However, there is a caveat. The combination of functions must make sense and must not overburden people manning the system if acceptance problems are not to hinder the realisation of the full potential of CIM technologies.[15]

As work processes are progressively integrated and many specialised jobs are abolished because of the reduced division of labour, work organisation tends towards the group pattern. Relatively autonomous groups are organised and their members execute complementary tasks; they must be versatile enough to handle a variety of jobs in order to keep the system running smoothly, and have the ability to co-operate and communicate beyond narrow technical boundaries. They are also given a certain amount of autonomy in the choice of tasks and in planning their work. In this way existing qualifications can be used more efficiently and mutual coaching takes place. Such teamwork, if properly organised, results in greater job satisfaction.

There is a further argument for increasing the autonomy of workers at all levels: making CIM work requires initiative, alertness and considerable creativity on the part of all concerned. This can flourish only when people have sufficient autonomy in their work to apply their skills, knowledge and talents to challenging tasks and projects.

However, it is also true that workers need time to get used to increased responsibility and autonomy, which some may perceive as a mixed blessing. An absence of clear instructions or incomplete instructions can be frustrating and may make people insecure. It should be obvious that increased responsibility cannot be sprung on workers without adequate preparation.

There is clearly more and more room for innovative work organisation; but, in the real world of manufacturing, including enterprises introducing CIM technologies, little progress in alternative forms of work organisation is evident. The inertia of

[15] U. Esser and A. Kemmner: "Arbeitssysteme für die erfolgreiche CIM-Einfuehrung", in Fir: *CIM* ..., op. cit., pp. 1-7.

existing organisational patterns appears to make the restructuring of organisations the most difficult part of CIM implementation. Divesting technical offices, for instance, of control functions and giving them to the shop-floor usually goes against the grain of the established power structure.[16]

Participative work organisation is, therefore, by no means an automatic outcome of introducing CIM. Management must consciously seek to overcome outdated, demotivating and unsuitable hierarchical forms of organisation, and this means shedding old power relationships, which is often a painful process fraught with pitfalls. Those who have a vested interest in maintaining the status quo are often in a strong position. In this context the results of a recent sample survey carried out in the Federal Republic of Germany deserve attention. Most enterprises continued to try to remove planning and control functions from the shop-floor through automation, concentrating them instead in the technical offices and at the supervisory level. Only a minority of enterprises attempted new organisational forms aimed at making systematic use of empirical shop-floor knowledge and competence by adapting the level of automation and reducing the division of labour.[17]

However, it should be some comfort to management deciding to base CIM on functional integration using human skills that this type of production is usually less capital-intensive than full-scale CIM since it is less computerised and requires less expensive software; the fact that many decisions are taken on the shop-floor also helps to make it flexible and to avoid machine down-time. As production planning, programming, machine setting and maintenance tasks are assumed by the group, process continuity is ensured. Moreover, existing qualifications of the workforce can

---

[16] C. Köhler et al.: "Alternativen der Gestaltung von Arbeits- und Personalstrukturen bei rechnerintegrierter Fertigung", in Kernforschungszentrum Karlsruhe: *Strategische Optionen der Organisations- und Personalentwicklung bei CIM*, Forschungsbericht KfK-PFT 148 (Karlsruhe, Aug. 1989), pp. 61-118.

[17] Schultz-Wild et al., op. cit., p. 194.

normally be used and few new ones are required. There is also a reduction of throughput time.[18]

Concern has been expressed that CIM will lead to the social isolation of the relatively few workers remaining on the shop-floor to mind the system. The fact that more and more communication takes place via computer terminals and opportunities for social contact are diminished may well have an adverse effect on individuals and the working atmosphere in plants. Ultimately such dissatisfaction could have negative consequences for the production process overall. At any rate, system designers need to keep this aspect in mind and provide opportunities for social contact as a contribution to the quality of working life.

Since the introduction of CIM technology in industry is in its infancy, much of what can be said about its impact on work organisation and working conditions must necessarily remain conjecture at the present stage. However, the opinion of experts is a valuable indicator of future trends. A survey conducted by experts in the Federal Republic of Germany in 1987 shows an interesting pattern of changes following the introduction of CIM technologies (figure 9).

Hazards in the new working environment

Although the new job requirements in CIM systems are gradually becoming better known, there is still much uncertainty about the new occupational safety and health hazards they may pose. It stands to reason that physical risks are diminished because fewer workers are in direct contact with production equipment and most production takes place without direct human intervention. However, the pace of work is usually faster and the amount of shift work and overtime greater; all these factors tend to increase fatigue and the risk

[18] Brödner, op. cit.

Figure 9. Changes in work organisation and working conditions (percentages)

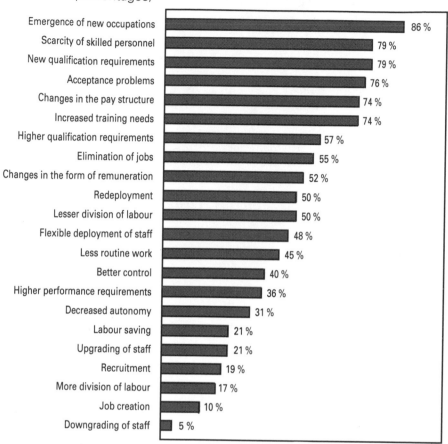

Emergence of new occupations	86 %
Scarcity of skilled personnel	79 %
New qualification requirements	79 %
Acceptance problems	76 %
Changes in the pay structure	74 %
Increased training needs	74 %
Higher qualification requirements	57 %
Elimination of jobs	55 %
Changes in the form of remuneration	52 %
Redeployment	50 %
Lesser division of labour	50 %
Flexible deployment of staff	48 %
Less routine work	45 %
Better control	40 %
Higher performance requirements	36 %
Decreased autonomy	31 %
Labour saving	21 %
Upgrading of staff	21 %
Recruitment	19 %
More division of labour	17 %
Job creation	10 %
Downgrading of staff	5 %

Source: Köhl, Esser and Kemmner, op. cit., p. 131.

of accidents. It has been found that work at computer terminals can be very stressful, particularly in the case of CAD, and it is not unusual for designers or draughtsmen to work for six to eight hours a day at CAD workstations. Similar work intensity has been observed in other computerised occupations. The computer, which was designed to facilitate work, ends up exhausting its users. In the long term this can lead to negative health effects.

The findings of a comparative ILO study on CAD applications are instructive. In general, CAD users experienced high job satisfaction because their work was more challenging than traditional draughting. Nevertheless, the report cautions:

... CAD users have also remarked that screen work has something of a "hypnotic" effect which leads to mental exhaustion. Some say that they feel completely drained after working for long hours with a system that exerts a kind of "horrible fascination" as the dialogue with the computer drives the operator on at a fast pace ...

Display terminal work can result in eye strain and back and neck problems. However, the extent of such complaints depends on whether ergonomic factors were taken into account in the design of the equipment and on the amount of time spent at the terminals. In this respect much progress has been made in recent years and screen displays, particularly in colour, now tend to cause less eye strain. Modern workstations are usually well designed and easy to operate.[19]

A potentially very serious problem is that an increasing number of psychosomatic illnesses appear to be caused by work with the new automated systems. Workers confronted with the expensive and complex equipment often do not feel up to the task assigned to them and have a sense of being powerless to intervene in the production process though they are responsible for running it. The combination of great responsibility and insufficient qualifications to master the job at hand or to intervene is extremely stressful. The stress may be aggravated by frequent breakdowns which have to be repaired quickly. An excessive workload also acts as a barrier to the acquisition of new qualifications. A state of constant stress can lead to nervous and physical disorders and is said to affect a disproportionate number of workers in advanced manufacturing systems. While training and ergonomically designed workplaces may help solve the problem, it can be averted more easily if system designers ensure from the outset that excessive demands are not placed on system users and maintenance staff. At the same time, however, they should not go to the opposite extreme and make jobs undemanding and monotonous, a factor that also causes stress.

[19] Ebel and Ulrich: "Some workplace effects of CAD and CAM", op. cit., p. 363.

Another cause for concern is the fact that work with CIM is more sedentary than traditional production work and requires more brainpower than muscle. As machines and robots take over materials and components handling, physical activity is greatly reduced and this too can be a serious threat to health. Countermeasures may well be needed.

A further cause of fatigue and stress is the poor design of much computer software. Much of it is not user-friendly and is ill-adapted to actual workplace requirements. This can make the interaction between workers and machines very cumbersome. "Cognitive" ergonomics addresses these problems. However, it is a relatively new science and improvements in software design that take into account research findings are only gradually forthcoming.

Another area of continuing concern is the impact of industrial robotics, an important component of CIM, on safety and health. There is wide agreement supported by case studies and accident statistics that the use of robots has significant beneficial effects as it eliminates much unpleasant, monotonous, repetitive, fatiguing, dangerous and heavy work. In particular, it improves safety in such environmental conditions as cold, heat, intense light, darkness, radiation, vibration, noise, toxic emissions, fire hazards, isolation of workers and inaccessibility of workstations. Nevertheless, like all machinery, robots can have harmful effects unless handled with the necessary caution. Sources of hazards include errors in the software of the control system and electrical interference from the mains or from radiation. The electrical, hydraulic or pneumatic controls may malfunction, causing unexpected results. It appears that the majority of recorded accidents with robots have occurred during maintenance and repair work.

The ILO has recently reviewed the basic principles and practices in the safe use of robots applied in various countries.[20] The review contains a catalogue of recommended measures and technical

---

[20] ILO: *Safety in the use of industrial robots*, Occupational Safety and Health Series No. 60 (Geneva, 1989).

standards referring, among other things, to guarding, trip devices, reliability of control systems, diagnosis of malfunctions, maintenance rules, robot operation, and safety training and organisation.

Despite some overall improvements in the physical working environment, there is one highly automated industry in which physical health hazards have become a cause for concern: the semiconductor industry. Its initial reputation as a "clean" industry has faded. It operates, in fact, with a multitude of cancer-causing, mutagenic and highly toxic chemicals to which it is feared that workers are exposed on an increasing scale because of negligence, insufficient protection, lack of safety training and deficiencies in hazard control. Reports have brought to light a high incidence of headaches, respiratory diseases, nausea, skin rashes, allergic reactions and miscarriages among electronics workers. Moreover, the handling of microscopic components tends to lead to eye damage.[21]

Finally, even in today's most advanced manufacturing systems there are many operations which for technical or economic reasons cannot be automated. This is the case particularly of assembly tasks, where the work pace is frequently dictated by machines. Workers in such jobs are exposed to cumulative trauma disorders (CTD), i.e. soft-tissue injuries in tendons, muscles and nerves caused by repetitive motions which may lead to disabilities.

System designers, who usually have an exclusively technical or scientific background, tend to overlook such considerations when planning the installations. However, it is then that preventive measures have to be taken and environmental and ecological concerns addressed. The increase this implies in planning and investment costs is insignificant compared with the cost of rectifying ergonomic and environmental mistakes once a system is installed.[22]

The high work intensity and continuous pressure resulting from an ever-increasing volume and complexity of work in information

[21] "Silicon Valley not so safe for workers, CWA tells House", in *AFL-CIO News*, 8 July 1989, p. 3; T. Gassert: *Health hazards in electronics: A handbook*, Hong Kong, Asia Monitor Resource Centre, 1985).

[22] Wobbe, op. cit.

Computer-integrated manufacturing

technology jobs has often meant that staff are obliged to do without breaks and to work a large amount of overtime; the solution would, of course, be adequate staffing. Enterprises ought to have a strong interest in preserving the capacity and efficiency of their workforce over the long term through refusing to tolerate conditions of work that are detrimental to physical and mental health.

The principal objective here should be the creation of humane working conditions, for only workers who are treated first and foremost as responsible human beings will be prepared to commit themselves to company goals. A definition of humane work that is apposite to the new technology is the following:

Work is called humane if it does not damage the psycho-physical health of the worker, does not . . . impair his psycho-social well-being, meets his requirements and qualifications, allows him to exercise individual and/or collective control over working conditions and systems of work, and is able to contribute to the development of his personality in activating his potential and furthering his competences.[23]

The preparation of the workforce for CIM

If people are the key to successful CIM, obviously much hinges on their preparation for the new systems. In all industrialised countries there is a shortage of professional, technical and managerial personnel able and qualified to mastermind the implementation of CIM. This shortage may well be aggravated in the future by reductions in working time, a major demand of the metal trades unions in most industrialised countries. The lack of adequate computer hardware or software is not the only major constraint; at the shop-floor level also the necessary skills are mostly in short supply. The recruitment difficulties experienced by enterprises,

[23] T. Martin, E. Ulich and H.-J. Warnecke: "Appropriate automation for flexible manufacture", in R. Isermann (ed.): *Preprints*, Tenth World Congress on Automatic Control, Munich, July 1987 (Laxenburg, Austria, International Federation of Automatic Control, July 1987), Vol. 5, pp. 291-305.

despite much unemployment and the high initial salaries paid to capable young engineers and technicians in this field, are a case in point. This skill shortage may well explain many of the systems failures reported earlier. Often management does not appear to have a clear idea of where it is going, while workers' representatives are seldom aware of the intricacies and possible social consequences of CIM. There is clearly a case for taking particular care in recruiting capable and motivated staff and for making production work attractive to them.

CIM technologies are new and unfamiliar to workforce and management alike, and training institutions often lag behind the latest developments. However, this is not the only reason for the skill shortage. There is a good deal of evidence that employers have paid too little attention to further training of their workforce in new information technology skills. In the United States, for instance, it was found that in 84 per cent of plants using programmable machines the employers provided no systematic operator training for the new functions. Only machine vendors organised some training for their equipment.[24] Such reluctance of employers to impart costly training in such skills was particularly noticeable in periods of economic recession. Even in the Federal Republic of Germany, with its extensive further training network, it was observed that so far further training in CIM technologies was ad hoc and haphazard, particularly in small and medium-sized enterprises engaged in small batch production, which constitute about 90 per cent of all metalworking firms. Only very recently have efforts been undertaken to systematise training in this field and to determine specific qualification requirements and training needs. An interesting example is the new occupation, "CAM organiser", which has just been recognised by the competent authorities. A curriculum comprising 436 hours of practical and theoretical upgrading

[24] Kelley and Brooks: *The state of computerized automation in US manufacturing*, op. cit.

instruction in information technology subjects for qualified CNC machinists, including examination regulations, has been approved.[25]

There is no easy solution to the problem. But one way of tackling it is through systematic training and further training of the workforce based on a strategy specifically designed for the purpose and endorsed by management and workers' representatives. Such training is required before and during the installation of new equipment and should emphasise not only specialised technical competence, including computer literacy, but above all system knowledge, planning, organisational and communication skills, and group dynamics. Ideally, there should be a symbiosis of system and product design and training. Such an approach would also serve to familiarise the workforce with the new equipment and thus help to overcome anxiety. Such training needs to be carried out mainly by the enterprises themselves in co-operation with system suppliers, since CIM systems are tailor-made to the specific requirements of enterprises and training institutes rarely have the necessary expertise in leading-edge technology. If and when flexible automation becomes more generalised, it may well become necessary to reform training curricula at all levels and make existing training infrastructures more responsive to changing occupational requirements.

Another path lies in the widest possible use of expert advice at the planning stage of CIM and an open discussion of alternatives among all those concerned — including the workers' representatives who far too often find themselves faced with a *fait accompli*. A thorough discussion of the economic, technical, organisational and manpower requirements and of the objectives of a proposed innovation would facilitate an informed assessment of the social consequences and the negotiation of working conditions and training obligations. In introducing CIM both management and the

[25] Fraunhofer-Institut für Arbeitswirtschaft und Organisation: *Mikroelektronik und berufliche Bildung (Projektphase II): Erstellung einer Konzeption zur Weiterbildung in rechnerintegrierten Betrieben der Auftragsfertigung* (Stuttgart 1989); "CAM Organisator/CAM Organisatorin", in *IBV*, No. 29, 19 July 1989.

workforce are usually moving into uncharted territory and ought to recognise the fact.

Suppliers of technology and consulting firms specialising in system integration can play a key role in assisting enterprises in the introduction of CIM. Their contribution tends to be all the more crucial if the enterprise is small and inexperienced. Firms may need help with financial and strategic planning, feasibility studies, system planning, human resources development and project management (the co-ordination of different suppliers). While many of these services can of course be contracted, over-dependence on suppliers, such as computer firms, software houses, machine vendors and management and engineering consultants, has its drawbacks. It is in the interest of CIM users to remain in control and build in-house expertise.

In-plant further training for CIM, administered to an already highly qualified workforce, is claimed to be the most effective way of bridging skill gaps. At the same time, it must be recognised that the cost of such training can be prohibitive, particularly for small and medium-sized enterprises. There is, therefore, a need for collaboration between industry, employers' and workers' organisations, government, professional societies, and research and training institutions to ensure the funding and the organisation of such training. This is clearly the price that has to be paid for a more rapid diffusion of CIM technologies.

The impact of CIM on industrial relations

In the real world the transition to CIM systems, even when it is well planned and prepared, will rarely be accomplished without creating tension and conflict (whether open or concealed) in organisations. The workforce has good reason to be worried since there is ample evidence to show that its interests may not be taken sufficiently into account or may simply be neglected. Far too often technology is placed before people to cope with as they can, without having been properly trained to handle it or given a say in its choice. Responsibility is taken away from the workers and de-skilling occurs; the work is robbed of its meaning and becomes boring. Small wonder that systems fail.

Workers fear pay losses as a result of reduced overtime, redundancy, fewer promotion prospects and lower manning levels, machine pacing and intensification of work, expropriation of know-how through data-bases and expert systems, higher stress through responsibility for expensive capital goods, more intensive shift work, individual performance monitoring by the computer system, de-skilling and apprehension about the unknown, and the strain of having to adjust to new working patterns, including possible redeployment and the loss of rights acquired after a long period of struggle. This can be, but must not be, the outcome.

By applying a strategy that puts people first and seeks genuine consultation at all levels, such fears can be overcome. The positive aspects of introducing advanced systems and the new opportunities they offer will more readily be accepted; these include safer and less physically taxing jobs, enhanced learning and training opportunities, greater responsibility and more interesting assignments, better remuneration, more flexible working hours, generally better working conditions or greater job security in a more competitive enterprise. The use of new systems can help to overcome the mismatch between personal aspirations and actual work requirements, bring down the rate of absenteeism and enhance productivity without undue intensification of work. Innovations can, in fact, exert a beneficial influence on industrial relations if more emphasis is placed on consultation at all levels and less play is given to "the arrogant expertise of technologists".

In this respect the Informatics Society of the Federal Republic of Germany observes:

The machine perspective ignores the fact that the working person possesses strong, albeit tacit work motives whose fulfilment he or she looks on as a sort of existential necessity: the demand for self-reliant expertise and an unexposed scope of action, and endeavours to achieve an informational advantage (particularly over superiors). Every sort of preprogrammed work is apt to jeopardize fulfilment of precisely these motives and to produce resistance or evasive action. This is the essence of the acceptance problem.[26]

The positive aspects can carry the day only in an atmosphere of social dialogue and good will at the enterprise level. Adversarial industrial relations, coupled with arbitrary exercise of management power, could easily spell the failure of CIM projects. Their success presupposes reconciling the interests of management and workers and introducing flexibility in work rules, redeployment measures and adequate training. Dialogue between the social partners is thus essential for achieving product and process innovation, and higher productivity and flexibility in manufacturing.

Unfortunately, the technocentric approach to CIM is widespread and in fact entails an attempt by employers to reassert their control over the workforce. On the one hand, most managements resent workers' control over production systems and see it as an encroachment on their prerogatives, particularly in an adversarial climate of industrial relations. On the other hand, unions are often sceptical about greater involvement of workers in the enterprise and distrust managerial strategies encouraging identification with the company. In their view greater responsibility of workers for the production process blurs the lines between "them" and "us" and threatens union identity and solidarity.[27]

There is indeed evidence that in an adversarial climate of industrial relations management tends to resort to an excessive division of labour as a means of restricting the influence of trade unions. In such circumstances management tends to override union jurisdiction and avoids entrusting blue-collar workers with more autonomy and control (e.g. the programming of NC tools) in order to circumvent rules and constraints imposed by collective bargaining agreements. This situation has the perverse effect that unionisation –

[26] IFAC: "Computers and responsibility . . .", op. cit., p. 12.

[27] Turnbull, op. cit.

Computer-integrated manufacturing

whenever it leads to restrictive and inflexible work rules and job control, such as manning requirements and seniority rules enforced by the unions – inhibits skill acquisition by blue-collar workers and their upgrading.[28] The blue-collar unions are in any case in a difficult situation because CIM technology is bound to accelerate the decline of their influence as a result of membership losses caused by the falling proportion of unskilled and semi-skilled workers in advanced manufacturing systems.

However, the greatest threat to workers' autonomy and to job satisfaction stems from centralised control, accompanied by demotivating rigidity and formal procedures in the production process. Decentralised systems coupled with a maximum of decision-making on the shop-floor are well suited to small batch or customised production, tend to enrich jobs and qualifications, reduce machine down-time through flexibility in work assignments, better scheduling and maintenance and, therefore, enhance productivity. They often prove to be economically superior to rigidly centralised systems with an excessive division of labour.[29] These factors should be taken into account when collective agreements are drawn up.

A large degree of consensus and co-operation is indeed necessary if CIM systems are to work smoothly, though this does not exclude a resolute defence of the workers' rights and interests. It must not be forgotten that highly skilled workers and technicians and their representatives in integrated manufacturing are in a strong position and cannot easily be replaced. Enterprises installing CIM depend on the quality and commitment of their workforce; qualified personnel are needed to maintain the complex and costly equipment and keep the system working. Moreover, advanced manufacturing systems are vulnerable to strikes, or work to rule, by a small proportion of their

---

[28] M. R. Kelley: "Unionisation and job design under programmable automation", in *Industrial Relations* (Berkeley, California), Spring 1989.

[29] F. Manske: "Social and economic aspects of alternative computer-aided production systems in small and medium batch runs", in IFAC: *Design of work in automated manufacturing systems*, Proceedings of IFAC Workshop, Karlsruhe, Federal Republic of Germany, 7-9 November 1983 (Oxford, Pergamon Press, 1984), pp. 51-55.

workforce, and responsible management will therefore be well advised to seek the social dialogue and collective agreements needed to provide a proper framework for their operation. An unorganised workforce kept in check by management prerogatives and arbitrariness, subdued by authoritarian supervision and anti-union policies, could easily jeopardise the success of CIM. Industrial relations based on mutual confidence and respect will be far more conducive to success.[30]

A telling example of the vulnerability of modern integrated plants to industrial action was reported in December 1986 by *Business Week*.[31] The highly robotised Waterloo tractor plant of John Deere and Co. was paralysed for a considerable time by a dispute with the United Autoworkers' Union which claimed redundancy protection in a new collective agreement. At the Waterloo plant it was impossible to shut down one part of the operation without bringing the rest of the plant to a standstill. This induced the company to rearrange the plant into smaller and more flexible "islands of automation".

The possible undermining of control systems and management objectives by a discontented workforce in the advanced technological environment has been described in these terms:

All [management control systems] have proved susceptible to what might be termed worker subversion. Once subordinates have come to understand the operation of the control system, they may, through superior operating understanding and sheer ingenuity, find the way to weaken its force and even, as in the "capture" of payment-by-results systems, turn it against managerial objectives. . . . If computerisation promises to provide management with a wealth of control-relevant information about the performance of subordinates, there may be little to guarantee that the information itself will be correct. A little distortion goes a long way in the interdependent sets of data stored by computers. Employees may falsify the information which they key in, whether to improve their own compensation or to ease the burden of hierarchical control. . . . Computer utilisation induces some people, and particularly creative people whose jobs

---

[30] F.-J. Kador: "Das Soziale in High-tech-Unternehmen", in *Der Arbeitgeber* (Cologne), No. 3/40, 1988, pp. 94-95.

[31] "Thinking ahead got Deere in big trouble", in *Business Week* (New York), 8 Dec. 1986.

underuse their skills, to play with the system. Computerised systems may be even more readily undermined and sabotaged by disgruntled employees than is the assembly line, which was supposed to chain the worker to the job but which proved intermittently vulnerable to sabotage and shutdowns.[32]

However, it should also be borne in mind that as hierarchical structures change and as middle management is threatened by CIM the role of unions and workers' representatives in enterprises may be weakened. Autonomous groups of highly qualified staff may be able to exert a more direct influence on the determination of their working conditions and thus may feel less need for union representation and intermediaries in their dealings with management. As the minority of core workers come to depend less on union intermediaries, the unions will find it increasingly difficult to organise the underprivileged workers on the periphery such as less-skilled part-time workers, workers in small subcontracting firms and others in precarious employment conditions. With the more widespread introduction of CIM technologies, the position of the unions can be expected to weaken.

Since unions tend to recognise the need for enterprises to be competitive, they have generally raised little opposition to technological change, and have often shown a positive attitude. Nevertheless, there have been situations when conflicts among unions arose because CIM technologies cut across narrow traditional craft demarcation lines and challenged established work practices. Such conflicts were not conducive to the adoption of advanced technologies. In the past decade many unions have been involved in technology bargaining and new technology agreements have often eased the transition to advanced technology, as the basis of an understanding between management and workers' representatives. However, such bargaining is often hampered by an understandable lack of technical expertise on the part of union negotiators. Unions have an interest in training their representatives and building

[32] P. Ryan: "New technology and human resources", Part II of G. Eliasson and P. Ryan: *The human factor in economic and technological change*, OECD Educational Monographs No. 3 (Paris, OECD, 1987), pp. 115-116.

sufficient expertise in their ranks to deal with the new social issues raised by the introduction of CIM; some have, in fact, issued guide-lines for the conclusion of such agreements to assist their rank and file in negotiations with management.

Given the flexible nature of CIM technologies, there will need to be more recourse to decentralised plant bargaining than to industry-wide bargaining since many issues are raised that can only be resolved at plant level. In enterprises introducing CIM, the social dialogue obviously cannot be limited to questions of remuneration and benefits. At any rate, payment-by-results systems may well have to be redesigned since the success of the system and higher productivity depend essentially on the built-in quality assurance system and reduction of down-time of the automated equipment, and not on the output of individual machine operators. Higher skills and system improvements must be rewarded. The dialogue will have to embrace questions related to the implementation of the new technology, such as more flexible working-time arrangements, adjusting working conditions to teamwork, no-redundancy clauses, redeployment, manning levels, skill upgrading and social plans in the event of plant closures.

Since technology agreements could ease the tensions associated with the introduction of CIM, it is useful to recall the main content of such agreements. In an ILO overview it was found that the following issues were usually covered:

— advance notice and information disclosure to workers and/or their representatives;
— a requirement for consultation and negotiation on any or all of the aspects of introducing new technology, including the establishment of procedures for such consultation or negotiation;
— access to outside expertise by workers and their representatives, or by joint committees, to help them play a full part in the procedures leading to the introduction of new technology, including consultation or negotiation;
— training for workers' representatives;

- training or retraining for workers, especially those whose jobs are eliminated or changed;
- proposals for work organisation and job design negotiations and for discussion on job content and job satisfaction;
- monitoring of workers' performance with new technology;
- negotiation over job grading and evaluation procedures, where jobs are created or changed by the introduction of new technology;
- the sharing of benefits arising from the use of new technology, such as higher pay or shorter working hours;
- provision of rest pauses and maximum hours to be worked in front of visual display units or other equipment;
- negotiation on working time and related issues, such as reduction or rearrangement of weekly working hours, or shift work associated with the introduction of new technology; and
- occupational safety and health and ergonomics.[33]

[33] Compiled from *Social and Labour Bulletin* (Geneva, ILO).

THE HUMAN AND SOCIAL PERSPECTIVE

The indispensable human factor

If we accept that a technocentric approach to CIM would be inefficient and counterproductive, it follows that we must recognise the key role of the human factor. The difficult part is to assign a really effective function to the people involved, to help them master the production process and utilise to the full their knowledge, capabilities and skills. This means providing appropriately designed jobs and workplaces, creating adequate interfaces between workers and machines, making the production process transparent, setting up suitable forms of work organisation and providing the necessary training for workers.

It is important to look at the potential weaknesses and strengths of the human factor in a CIM environment. Human beings involved in the production process are apt to make mistakes, particularly when they are under physical or psychological stress. Noise or bad lighting can lead to fatigue. The rate of error also increases with information overload. Survey findings show that 70-90 per cent of the failures of technical systems are due to faulty human intervention or system design. Human beings do not always concentrate fully on their work:

they sometimes come to wrong conclusions or fail to act when they should. Their behaviour at work can be unpredictable. Is this sufficient reason to banish people from the production process?

There is clearly a wide range of tasks and functions that are best done by machines, industrial robots and computers. This range is being constantly extended as more and more manual operations are taken over by machines. The improvement of sensors and actuators makes robots and other production equipment more versatile, rapid and exact than human beings. Some jobs can be done far more efficiently and reliably by information systems and computers. This is particularly the case for routine functions such as data collection and statistical analysis, as well as many surveying and control functions that serve as the basis for automatic process and quality control in production. Hence the execution of an increasing number of specialised functions can no doubt be transferred to machines and computers. We are therefore faced with a growing complexity of such technical systems.

As such systems become more complex, however, they also tend to be more fail-safe. The dilemma has been described as follows:

... every conceivable effort must be made to avoid avoidable errors; but computer science is only gradually learning from other less complex technical systems that there is — and can be — no such thing as faultless hardware, error-free operating systems and large programs that are totally correct. This fact must be taken into account as early as the design stage, since the utterly unwarrantable illusion that it is possible to build an error-free system usually results in the system's being unable to cope with unexpected errors in its subsequent uncontrolled breakdown. More serious, though, than the inability to build error-free systems is the fact that, despite significant advances in this field, there are still no reliable test methods available. Although software developers are well aware of these problems and the fundamental impossibility of completely eliminating systems' susceptibility to error, insufficient effort is made to publicly correct the image of the perfectly functioning computer system which is frequently fostered in advertising.[1]

In fact, technical systems break down frequently, and the cost of such breakdowns is high. They can be perfected or repaired but this requires skilled human intervention and means that the workers

[1] IFAC: "Computers and responsibility...", op. cit., p. 3.

Computer-integrated manufacturing

responsible for the operation have to make choices and decisions which no technical system can make for them. Quick intervention based on experience and a knowledge of the system's limits is often required. Despite their shortcomings, human beings are thus indispensable for an optimal and efficient use of automated equipment. The qualified, motivated and experienced worker familiar with the system can cope with uncertainty and assess situations, find and interpret faults rapidly and correct them. Judgement backed up by technical knowledge and experience, understanding of the system and common sense is a human quality that cannot be replaced by computers or artificial intelligence in the foreseeable future. In CIM systems, although machines and computers may well take over most routine and physical tasks, they do not relieve the people involved from thinking, critical decision-making and responsibility.

The design of CIM systems

Since the human factor cannot be replaced it is essential to design and plan CIM systems in such a way that those working on them can do their job in optimal conditions and can effectively apply their empirical knowledge. This means, first of all, that they must not be made totally and helplessly dependent on the system. Such dependence could have serious consequences when system errors cannot be corrected in good time. It also limits initiative, improvisation and creativity, and fosters blind reliance on routine which in turn makes the systems, and consequently the enterprise, vulnerable. Overdependence on systems and machines has not only caused disasters in nuclear and chemical industries, but has also produced failures in automated production which are perhaps less spectacular, but none the less very costly.

Workers employed in "human centred" CIM systems should be able to intervene in the production process in order to optimise it. This means that the system must allow shop-floor programming of CNC equipment on the basis of indications provided by and discussed with the design office. The implementation of such organisational principles presupposes the availability of appropriate worker/machine interfaces and software. An example is a portable electronic sketch pad which designers can use to discuss their ideas with shop-floor personnel thus helping to overcome their notorious divorce — accentuated by computer-aided design — from the realities and constraints of the production process.[2]

The considerable range of job design for CNC machines, from simple to more complex alternatives, as is shown in figure 10. Much depends, of course, on the qualifications and versatility of workers.

Systems development will need to concentrate on devising planning and organisational techniques that effectively help to overcome the division of labour and foster the integration of functions, as well as the decentralisation of decision-making and detailed planning.

If human interaction with CIM systems is to be enhanced and optimal performance achieved, user-friendly tools for such functions as rerouting production flow, reprogramming or equipment repair must be developed in order to increase the decision-making power of production workers. This is feasible as the design of a decision support system for lathe operators/machinists has demonstrated. It has the following characteristics:

1. Performs all detailed and routine calculations, but leaves the machinist to use his machining experience, skill and knowledge of the characteristics of workpiece material, cutting tools and machine tool, to exercise judgement over the results of the calculations (corresponding design guide-lines: complementarity and operator control).

[2] Paper presented by M. Cooley to a conference on vocational training as an investment in the future, Essen, 2-3 May 1988, cited in *BFZ Info* (Essen), 1988, No. 3, pp. 4-5.

Figure 10. Alternative jobs for CNC machine operators

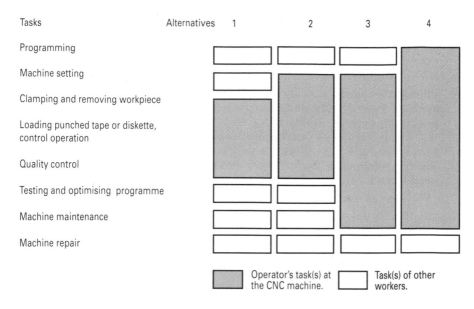

| Tasks | Alternatives | 1 | 2 | 3 | 4 |

Programming

Machine setting

Clamping and removing workpiece

Loading punched tape or diskette, control operation

Quality control

Testing and optimising programme

Machine maintenance

Machine repair

Operator's task(s) at the CNC machine.   Task(s) of other workers.

Source: Gesamtmetall: *Mensch und Arbeit* (Cologne, Edition Agrippa, 1989), p. 11.

2. Gives the machinist the freedom to accept or reject computer-suggested cutting parameters, but does not constrain his own choice to the range of data available, nor allow his choice to be restricted by the constraints forming part of the mathematical model of the machining process (corresponding design guide-lines: operator control, disturbance control, fallibility and operating flexibility).

3. Allows the machinist to determine the degree of computer support available and also the way in which that computer support is used (corresponding design and guide-lines: operator control, interactivity and operating flexibility).

4. Consists of a relatively straightforward procedure, related to shop-floor knowledge as much as possible, for calculating the cutting parameters (corresponding design guide-lines: transparency and accountability).

5. Provides the machinist with facilities to adjust those system parameters having an important influence upon the calculations (corresponding design guide-lines: operator control, disturbance control, fallibility and operating flexibility).

6. Presents results in a form that can be easily assimilated (corresponding design guide-lines: interactivity, transparency, compatibility, accountability, minimum shock and error reversibility).

7. Allows the machinist to examine the state of the mathematical model of the cutting process in relation to various combinations of cutting parameters (corresponding design guide-lines: operator control, interactivity, transparency, accountability, minimum shock and error reversibility).
8. Requires the machinist to take controlling action to initiate the machining process (corresponding design guide-lines: operator control, minimum shock, error reversibility and fallibility).
9. Informs the machinist of the consequences of his decisions in relation to the model of the machining process in the software, allowing him to compare these predictions with the reality in which he operates, but at the same time not restraining him from carrying out those decisions if he wishes to do so (corresponding design guide-lines: operator control, transparency, accountability, minimum shock, error reversibility, fallibility and operating flexibility).
10. Provides the machinist with a choice of working methods, thus allowing him to develop his skills in selecting an appropriate method of working and allowing him to explore and to learn different techniques (corresponding design guide-lines: operator control and operating flexibility).[3]

Another problem that has to be borne in mind is that there tends to be a greater physical distance between the personnel operating or supervising the equipment and the production process itself. Much visual and manual control has been replaced by sensors that transmit data to screens and data bases. The worker is faced with control data at the workstation, but loses direct contact with the production process which can often only be monitored from a control room. It has been found that such distances from the process may make quick reaction and the correction or compensation of system faults more difficult because warnings emitted by the system can be misinterpreted or neglected and workers lose the "feeling" for the process. They may even lose touch with the consequences of their own actions.

This means that the process must be designed transparently so that it is comprehensible and sufficiently accessible to the worker, without creating hazards, in order to allow the required intervention

[3] P. T. Kidd: "The social shaping of technology: The case of a CNC lathe", in *Behaviour and Information Technology* (Basingstoke, Hampshire), Vol. 7, No. 2, 1988, pp. 200-201.

to be made. Some research conducted on the empirical knowledge of machines and materials that experienced skilled workers possess indicates that they develop a feeling, almost a sixth sense, that tells them what is wrong with a machine and how it works best. This enables them to correct deviations from the normal production process and to improvise when problems arise. This capability is precious and its importance for the smooth running of advanced manufacturing systems should not be underestimated.[4] The conclusion to be drawn is that CIM programmes should follow as far as possible the thought processes of machinists and technicians, and not try to implement abstract engineering concepts.

In the technocentric approach to CIM there is clearly a tendency to incorporate most production knowledge in the computer and expert systems without giving workers sufficient opportunity to exercise their skills — skills that may then waste away from lack of use. At the same time, empirical knowledge can no longer be acquired. This can make production systems very unwieldy and vulnerable and their rigidity would cause considerable constraints. It can also erode the human knowledge base of the enterprise to such an extent as to put its future in jeopardy. This is too high a price to pay for enterprises that depend on the qualifications, skill and adaptability of their workforce.

The technocentric approach aims at eliminating human intervention in the production process as much as possible. The consequences have been pointed out by the Informatics Society of the Federal Republic of Germany:

The ideal of this view of technology is the fully automated system. Existing information technology tends to adopt technical solutions which are pledged to this ideal. Here, information-processing systems are not designed to support the human beings working with them; instead, the human beings are forced to submit to the technical and organizational logic of the automated process. The "machine-like nature" of this view of development implies an inversion of the relationship between the subject and object of the work process. This definition of

[4] F. Böhle and B. Milkau: *Vom Handrad zum Bildschirm: Eine Untersuchung zur sinnlichen Erfahrung im Arbeitsprozess* (Munich, Campus Verlag, 1988).

the subject/object relationship entails the risk that the computer no longer assumes a supportive function, but a dominant one. Instead of working *with* the (tool) computer, the human being works *at* the computer. The degradation of the working person from the subject of the work action to machine operator soon reduces his or her willingness (or even ability) to act responsibly at work.

. . .

Organizers who base their ideas on Tayloristic notions view the social organization of a business enterprise as a controllable machine in the sense of classical mechanics. Hence, they expect information systems to enable them, by applying watertight mathematical logic, to make all work procedures transparent and to link them up with one another. The human being (seen here as an "information-processing system") is an element included in planning and fitted into the system as a whole in accordance with "preprogrammed task structuring". Such an approach to the organisational application of information technology proves illusory.[5]

And another word of caution:

work may indeed be routinised by building discretion into the computer at the expense of that exercised by the operator. But the process is subject to strong constraints in practice. Even apparently routine jobs often turn out to require human ingenuity and initiative. While there is no logical impossibility to embodying all eventualities in a computer programme, practical possibility is a different matter.[6]

There are, indeed, signs that the technocentric approach may lead to the disaffection of the workforce. Research findings in the United States suggest that workers in high-technology industries are less satisfied than other manufacturing workers because of more rigid rules, stricter discipline and closer supervision and monitoring.[7] Such disaffection has been noted particularly in the case of production workers who feel threatened by de-skilling. It was found that in 1987 only one out of seven blue-collar workers who operated computer-controlled machines had been given the opportunity to exercise major responsibility for programming.[8] Management also has a tendency to leave the trouble-shooting and maintenance and repair of advanced systems to specialised services without involving

[5] IFAC: "Computers and responsibility . . .", op. cit., pp. 5 and 12.

[6] Ryan, op. cit., p. 112.

[7] Hodson and Hagan, op. cit.

[8] Kelley and Brooks, op. cit.

the operators of the equipment, and this neglect of shop-floor skills has led to a great increase in production down-time.[9] The obvious solution is to entrust as much responsibility for maintenance as possible to suitably qualified workers on the shop-floor. People operating CIM systems must be in a position to dominate them, and not vice versa.

The main requirement — that human beings should be empowered to use the technical systems as a tool — implies a human-centred approach to system design that is often overlooked. This imperative has been expressed as follows:

> The demand for conformity with the tool analogy is, however, neglected in system development if human beings are viewed — like computers — as information-processing systems. This — regrettably widespread — view tends to conceal human qualities and skills. If we are to develop information-processing "tools" for human beings, the machine-oriented view of the human being needs thoroughly revising. The tool analogy used in connection with the new information technology must include here consideration of the quality of the overall relationship between human beings and machines.[10]

The practical problems of taking all these aspects into account in designing CIM systems and of making optimal use of the human factor must not be underestimated even when such human-centred systems are recognised as superior. Though by no means easy to organise, a multidisciplinary approach is needed: managers and engineers responsible for the system must be committed to such an approach and should involve ergonomists, training specialists and social scientists. There are few tried methods of proceeding since the technocentric approach, which neglects ergonomics and social concerns, has prevailed up to now. Moreover, engineers and ergonomists often tend to be at cross purposes. This is certainly a field that requires further study.[11]

[9] P. Chabert and P. Laperrousaz: "Modernisation — Attention aux idées reçues!", in *L'Usine nouvelle* (Paris), No. 3, 21 Jan. 1988, pp. 4-8.

[10] IFAC: "Computers and responsibility . . .", op. cit., p. 7.

[11] J. M. Corbett: "Human-centred advanced manufacturing systems: From rhetoric to reality", in *INFO Pack No. 4* (Karlsruhe, Committee on Social Effects of Automation, IFAC, Mar. 1989); also to be published in *International Journal of Industrial Ergonomics* (Amsterdam), 1990.

The main responsibility remains with the engineers who design the systems and software. They far too easily lose sight of the human and social dimensions of systems or even fail to recognise that these exist. The social responsibility of the system designers cannot be emphasised enough as they have a direct influence on the quality of working life. They should, therefore, be trained to develop technologies with desirable social outcomes in mind. Social aims in systems design may well include the opportunity to develop and use skills, the control of stress and the opportunity for interaction between workers on the job.[12]

[12] P. T. Kidd: "Technology and engineering design: Shaping a better future or repeating the mistakes of the past?", in *IEE Proceedings*, Vol. 135 (Part A), No. 5, May 1988, pp. 297-302.

RESEARCH ON THE PROMOTION OF THE HUMAN FACTOR IN CIM SYSTEMS

I t is one thing to identify the desirable features of a CIM system that takes into account the human factor, but quite another to design such systems so that they work in practice with an adequate rate of return and are acceptable to management. In most manufacturing enterprises the division of labour is still entrenched and tends to be used as a means of social control. The Taylorist mentality in production management is widespread and resistant and will not disappear from one day to the next, particularly in mass production. A conscious effort is therefore needed to promote the human-centred approach to CIM. As the movement gains ground a great variety of research projects have been launched throughout the industrialised world with the objective of developing models and defining the technical, ergonomic, organisational, social and training criteria for such systems.[1] This research forms part of several national programmes, and international programmes of the European Community[2] (e.g. FAST, ESPRIT, EUREKA, RACE, COMETT, BRITE), involving a wide range of technical, social science and training research institutes, enterprises and employers' and workers' organisations.

Partial results have been reported; some are encouraging, others less so. Contradictory findings are not uncommon. It is clearly not

[1] The present state of research is reflected in: Union-Druckerei und Verlagsanstalt: *CIM oder die Zukunft der Arbeit in rechnerintegrierten Fabrikstrukturen: Ergebnisse einer Fachtagung der IG-Metall* (Frankfurt am Main, 1987).

[2] European Communities: *Europe and the new technologies* (Brussels, 1988).

easy for work science, ergonomics, and design and systems engineering to come to grips with the manifold aspects and the complexity of the problems at hand. Interpretations, recommendations and proposed solutions also depend considerably on the point of view and the ideological leanings or ethical principles of the researchers; there are few certainties but much wishful thinking. It is obviously too early to assess the impact of such research on industrial practice. However, as the technocentric approach to CIM runs into growing difficulties the opportunities for using alternative models are bound to increase as enterprises may be more willing to give them a chance.

There are some hopeful signs. In the United States the tripartite Work in America Institute, recognising the decisive influence of the human factor on productivity, co-operates with large enterprises in an effort to enhance its role. Only a small minority of enterprises are involved, but it is a beginning.[3] The Massachusetts Centre for Applied Techology in Boston works in the same direction.

In France the Ministry of Research and Technology is sponsoring a research project entitled "The enterprise faced with integration (CIM)", in which various university institutes are involved. It places emphasis on new forms of factory organisation to make work more meaningful and on structuring the information flow in enterprises.

A number of European Community projects are also worth mentioning in some detail. Two projects concerned with human factors have been funded within the CIM research area of ESPRIT: No. 534, "Development of a flexible automated assembly cell and associated human factors study", and No. 1199, "Human-centred CIM systems".

Project No. 534, begun in January 1985, was scheduled to last for five years. Unfortunately, owing to the withdrawal of the prime contractor, the project was terminated prematurely after only two years. Its aim was to design and develop a prototype automated

---

[3] "Réseau de 20 entreprises appliquant des systèmes socio-techniques", in *UIMM Social International* (Paris), Feb. 1989, p. 11.

flexible assembly cell for the manufacture of mechanical assemblies with a volume of up to 0.5 cubic metres and weighing up to 30 kg, in low-batch quantities (ideally as low as one). The project also included a study of the human factors related to work design. This aimed to produce a set of generalised design rules for human factors in CIM application, covering such topics as allocation of functions between worker and machine, work organisation, health and safety, hardware and software ergonomics and environmental ergonomics.

Project No. 1199 started in 1986 and was completed in 1989. The objective of the project was to develop prototype manufacturing systems comprising integrated computer-aided design, manufacturing and planning packages in which the roles of the human operators were optimised; those working with the system would be made responsible for all tasks, and computerised systems would be used as a back-up. The project was to demonstrate that a system that allows for the use and development of human skills and abilities within a CIM environment can be more effective, more robust and more economic than the conventional total automation approach adopted by many enterprises.

A number of applications of the various systems developed by this project have been, or will be, undertaken. At one user site a turning cell consisting of two CNC lathes and a work-handling device have been installed. The machinists working in this cell have responsibility for the programming of the machine tools, scheduling of the work and quality control.

At another manufacturing site, the proposal is to reorganise the whole manufacturing layout. The factory, which is at present centrally controlled with a functional layout, will be reorganised into production islands, each capable of manufacturing a wide range of products and with all the machines (with one expensive exception) in sufficient quantity to do so. The team of people working in each island will be responsible for almost everything that goes on there: quality control, programming of machines, boundary management, scheduling of the machines in the island, minor maintenance and repairs, stock control, and so on.

A prototype production island with supporting computer-based tools has been developed. It has been tested in a laboratory environment and was successfully demonstrated in a real production situation during a one-week period in 1989. Plans are being made for the implementation of the system in a number of factories.

Mention should also be made of the promotional activities carried out in this field by the International Federation of Automatic Control (IFAC), whose Committee on Social Effects of Automation makes a considerable effort to assess the social impact of control and automation technology and gathers concerned scientists, control engineers, managers, system designers, social scientists, industrial psychologists and specialists of other disciplines from the research institutes and industry of a wide range of countries. The Committee emphasises the social responsibility of control engineering and aims to establish socially desirable requirements for the development of automated systems, demonstrate design alternatives for such systems, strengthen links between technologists and social scientists, and disseminate knowledge of these topics among the scientific and technical community. It has encouraged a considerable number of projects in which industry has co-operated and has contributed in no small measure to the idea of the human-centred approach to CIM, that is slowly but surely making its way.[4]

Research looking more specifically at the role of the human factor in technical systems so far forms only a marginal part of all research into CIM. It is not the place here to discuss the research strategies of stakeholders in advanced manufacturing systems nor resource allocations by public and private bodies and industry. At the international level, this has been done extensively in recent years by the OECD.[5] However, if a more humane working environment, better conditions of work and a higher quality of life are to be achieved, ways and means must be found through more research devoted to this objective.

[4] IFAC: *IFAC Information: Aims-Structure-Activities* (Laxenburg, 1988 edition).

[5] G. Drilhon: "The research system under constraint", in *STI Review* (Paris, OECD), No. 5, Apr. 1989, pp. 129-162.

OUTLOOK

At present there is little chance of reconciling divergent views on CIM. Many of its advantages or faults are in the eye of the beholder. However, it is definitely not the panacea for all problems encountered in production that some seem to see in it. The promised land of total manufacturing integration is still far away, although an increasing number of enterprises appear to be engaged in an evolutionary process towards it.

By any standard, the introduction of CIM is a risky undertaking. If it is to be successful the firm's manufacturing organisation and product range will have to be reviewed and rationalised. The pace of transition will depend on the knowledge, qualifications and abilities of the planning and operating staff. The neglect of the human factor, the absence of systematic training and personnel planning and the maintenance of ossified and obsolete organisational patterns are fatal for the implementation of CIM.

CIM is certainly not a miraculous solution for enterprises falling victim to market forces, but if used as a strategic means to seize new opportunities, it may help them win back lost terrain. Enterprises are well advised to avoid great technology leaps that are liable to fail and to build up their technological capabilities systematically but gradually.

CIM is a leading-edge technology and its introduction requires long-term strategies, much research and development and, possibly, forgoing immediate financial benefits. The most essential element in such a strategy is the preparation of the workforce for the impending changes. This requires consultation at all levels and a systematic training effort. To neglect the further training of staff is inevitably very costly in terms of machine down-time and scrap production.

All experience gained so far speaks in favour of a cautious and incremental approach in order not to overstretch the assimilating and learning capacity of the workforce with the negative results which that implies. The fact remains that the trend towards more manufacturing integration is bound to continue and scientific advances will continue to offer solutions to outstanding technical problems. CIM is most likely to fail where it tries to supplant essential human qualities. The subjugation of people to machines and technical systems is proving more and more counterproductive. Instead, a type of work organisation is needed that enables and motivates people to use their theoretical and empirical knowledge and skills in mastering advanced means of production and operating them efficiently. CIM will only be as good as the people in charge of it.

This engages the social responsibility of the system designers; they should make the limits and the capabilities of technical systems explicit and design them as a tool in the service of the people operating them. A residual role for humans in the manufacturing process is dehumanising. CIM should result in an organisation that makes the best possible use of people's knowledge, skills, capabilities and talents, which are vastly underutilised. Moreover, system designers should not lose sight of the objective of creating humane working conditions and improving the quality of working life. The exposure of operators to excessive stress has become a major preoccupation in advanced manufacturing systems. More research needs to be directed to these objectives.

Are we heading in the wrong direction? Evidence suggests that the difficulties and complexities of introducing CIM on a large scale were initially underestimated. The technocentric approach aiming at the "unmanned factory" is now questioned for a very good reason: so far it has failed to produce the expected results. Its social and economic costs in terms of worker alienation, absenteeism and production losses have often been disregarded. This is having a sobering effect on the unconditional technocrats. There is probably not just one type of "factory of the future" but many alternative solutions to manufacturing problems. In fact, from the technical point of view there is a proliferation of choices.

There is little evidence of dramatic negative effects on employment. CIM will do away with unskilled jobs, but this will happen gradually since its diffusion is slow. At any rate, it will have only a marginal effect on overall employment levels in the foreseeable future. Moreover, job reduction is seldom a rationale for the use of CIM technologies because direct labour costs in modern manufacturing are in any case relatively low when compared with capital costs; such technologies are more likely to be used to economise on scarce skilled labour and overcome labour shortages. In particular, CIM opens up opportunities for an optimal use of expensive capital equipment through making operating times of factories progressively independent of the working time of the workforce for whom a wide array of flexible working schedules can be devised.

Will CIM really spell the end of Taylorism? It is definitely too early to pronounce its methods dead and buried. Taylorism will continue to subsist in mass production alongside dedicated automation and machinery, and so will the corresponding hierarchical management structures. However, mass production and market dominance of mass-produced goods are declining in many manufacturing activities. The markets demand differentiated, diversified and customised products, entailing a need for small-batch production. The flexible automation offered by CIM can, if properly conceived, do the job.

At times of accelerated technical change there is a need not only for innovative responses from both management and workers' representatives, taking into account technological and economic imperatives, but also for new opportunities to be at hand to improve the working environment and conditions of work. In the continuing adjustment process unions may need to shed inherited organisational structures and philosophies. However, the onus of making new work organisation effective is on management.

Integrated manufacturing systems are very vulnerable to disruption. Running them efficiently and, as far as possible, round the clock presupposes harmonious industrial relations, since work stoppages, go-slows or other types of resistance stemming from demotivating working conditions can cause major losses. The success of CIM, therefore, presupposes mutual understanding and co-operation between management and the workforce and its representatives. While the introduction of even well-designed CIM systems is bound to cause tensions, it also offers new opportunities for enhancing dialogue and breaking down barriers between the social partners – a chance not to be missed.

Computer-integrated manufacturing